Parallel Iterative Algorithms

From Sequential to Grid Computing

CHAPMAN & HALL/CRC
Numerical Analysis and Scientific Computing

Aims and scope:
Scientific computing and numerical analysis provide invaluable tools for the sciences and engineering. This series aims to capture new developments and summarize state-of-the-art methods over the whole spectrum of these fields. It will include a broad range of textbooks, monographs and handbooks. Volumes in theory, including discretisation techniques, numerical algorithms, multiscale techniques, parallel and distributed algorithms, as well as applications of these methods in multi-disciplinary fields, are welcome. The inclusion of concrete real-world examples is highly encouraged. This series is meant to appeal to students and researchers in mathematics, engineering and computational science.

Proposals for the series should be submitted to one of the series editors above or directly to:
CRC Press, Taylor & Francis Group
24-25 Blades Court
Deodar Road
London SW15 2NU
UK

Parallel Iterative Algorithms

From Sequential to Grid Computing

Jacques Mohcine Bahi
Sylvain Contassot-Vivier
Raphaël Couturier

CRC Press
Taylor & Francis Group
Boca Raton London New York

CRC Press is an imprint of the
Taylor & Francis Group, an **informa** business
A CHAPMAN & HALL BOOK

CRC Press
Taylor & Francis Group
6000 Broken Sound Parkway NW, Suite 300
Boca Raton, FL 33487-2742

First issued in paperback 2019

© 2008 by Taylor & Francis Group, LLC
CRC Press is an imprint of Taylor & Francis Group, an Informa business

No claim to original U.S. Government works

ISBN-13: 978-1-58488-808-6 (hbk)
ISBN-13: 978-0-367-38816-4 (pbk)

Library of Congress Cataloging-in-Publication Data

Bahi, Jacques M.
    Parallel iterative algorithms: from sequential to grid computing / authors, Jacques M. Bahi, Sylvain Contassot-Vivier, and Raphael Couturier.
        p. cm. -- (Chapman & Hall/CRC numerical analysis and scientific computing series)
    Includes bibliographical references and index.
    ISBN 978-1-58488-808-6 (alk. paper)
        1. Parallel processing (Electronic computers) 2. Parallel algorithms. 3. Computational grids (Computer systems) 4. Iterative methods (Mathematics) I. Contassot-Vivier, Sylvain. II. Couturier, Raphael. III. Title. IV. Series.

QA76.58.B37 2007
518'.26--dc22   2007038842

Visit the Taylor & Francis Web site at
http://www.taylorandfrancis.com

and the CRC Press Web site at
http://www.crcpress.com

Contents

List of Tables   ix

List of Figures   xi

Acknowledgments   xiii

Introduction   xv

1 Iterative Algorithms   1
   1.1 Basic theory   1
        1.1.1 Characteristic elements of a matrix   1
        1.1.2 Norms   2
   1.2 Sequential iterative algorithms   5
   1.3 A classical illustration example   8

2 Iterative Algorithms and Applications to Numerical Prob-
   lems   11
   2.1 Systems of linear equations   11
        2.1.1 Construction and convergence of linear iterative algo-
               rithms   11
        2.1.2 Speed of convergence of linear iterative algorithms . .   13
        2.1.3 Jacobi algorithm   15
        2.1.4 Gauss-Seidel algorithm   17
        2.1.5 Successive overrelaxation method   19
        2.1.6 Block versions of the previous algorithms   20
        2.1.7 Block tridiagonal matrices   22
        2.1.8 Minimization algorithms to solve linear systems   24
        2.1.9 Preconditioning   33
   2.2 Nonlinear equation systems   39
        2.2.1 Derivatives   40
        2.2.2 Newton method   41
        2.2.3 Convergence of the Newton method   43
   2.3 Exercises   45

3 Parallel Architectures and Iterative Algorithms   49
   3.1 Historical context   49
   3.2 Parallel architectures   51
        3.2.1 Classifications of the architectures   51
   3.3 Trends of used configurations   60
   3.4 Classification of parallel iterative algorithms   61
        3.4.1 Synchronous iterations - synchronous communications
               (SISC)   62

       3.4.2   Synchronous iterations - asynchronous communications
               (SIAC)   63
       3.4.3   Asynchronous iterations - asynchronous communications
               (AIAC)   64
       3.4.4   What PIA on what architecture?   68

4 Synchronous Iterations   71
   4.1 Parallel linear iterative algorithms for linear systems   71
       4.1.1   Block Jacobi and O'Leary and White multisplitting al-
               gorithms   71
       4.1.2   General multisplitting algorithms   76
   4.2 Nonlinear systems: parallel synchronous Newton-multisplitting
        algorithms   79
       4.2.1   Newton-Jacobi algorithms   79
       4.2.2   Newton-multisplitting algorithms   80
   4.3 Preconditioning   82
   4.4 Implementation   82
       4.4.1   Survey of synchronous algorithms with shared memory
               architecture   84
       4.4.2   Synchronous Jacobi algorithm   85
       4.4.3   Synchronous conjugate gradient algorithm   88
       4.4.4   Synchronous block Jacobi algorithm   88
       4.4.5   Synchronous multisplitting algorithm for solving linear
               systems   91
       4.4.6   Synchronous Newton-multisplitting algorithm 101
   4.5 Convergence detection 104
   4.6 Exercises 107

5 Asynchronous Iterations   111
   5.1 Advantages of asynchronous algorithms 112
   5.2 Mathematical model and convergence results 113
       5.2.1   The mathematical model of asynchronous algorithms . 113
       5.2.2   Some derived basic algorithms 115
       5.2.3   Convergence results of asynchronous algorithms 116
   5.3 Convergence situations 118
       5.3.1   The linear framework 118
       5.3.2   The nonlinear framework 120
   5.4 Parallel asynchronous multisplitting algorithms 120
       5.4.1   A general framework of asynchronous multisplitting meth-
               ods 121
       5.4.2   Asynchronous multisplitting algorithms for linear prob-
               lems 124
       5.4.3   Asynchronous multisplitting algorithms for nonlinear
               problems 125
   5.5 Coupling Newton and multisplitting algorithms 129

5.5.1 Newton-multisplitting algorithms: multisplitting algorithms as inner algorithms in the Newton method .. 129

5.5.2 Nonlinear multisplitting-Newton algorithms 131

5.6 Implementation 131

5.6.1 Some solutions to manage the communications using threads 133

5.6.2 Asynchronous Jacobi algorithm 135

5.6.3 Asynchronous block Jacobi algorithm 135

5.6.4 Asynchronous multisplitting algorithm for solving linear systems 138

5.6.5 Asynchronous Newton-multisplitting algorithm 140

5.6.6 Asynchronous multisplitting-Newton algorithm 142

5.7 Convergence detection 145

5.7.1 Decentralized convergence detection algorithm 145

5.8 Exercises 169

6 Programming Environments and Experimental Results   173

6.1 Implementation of AIAC algorithms with non-dedicated environments 174

6.1.1 Comparison of the environments 174

6.2 Two environments dedicated to asynchronous iterative algorithms 176

6.2.1 JACE 177

6.2.2 CRAC 180

6.3 Ratio between computation time and communication time . 186

6.4 Experiments in the context of linear systems 186

6.4.1 Context of experimentation 186

6.4.2 Comparison of local and distant executions 189

6.4.3 Impact of the computation amount 191

6.4.4 Larger experiments 192

6.4.5 Other experiments in the context of linear systems .. 193

6.5 Experiments in the context of partial differential equations using a finite difference scheme 196

Appendix   201

A-1 Diagonal dominance. Irreducible matrices 201

A-1.1 Z-matrices, M-matrices and H-matrices 202

A-1.2 Perron-Frobenius theorem 203

A-1.3 Sequences and sets 203

References   205

Index   215

List of Tables

5.1   Description of the variables used in Algorithm 5.7.   149

5.2   Description of the additional variables used in Algorithm 5.15.   163

6.1   Differences between the implementations (N is the number of processors).   175

6.2   Execution times of the multisplitting method coupled to different sequential solvers for a generated square matrix of size 10.10^6 with 70 machines in a local cluster (Sophia).   189

6.3   Execution times of the multisplitting method coupled to different sequential solvers for a generated square matrix of size 10.10^6 with 70 machines located in 3 sites (30 in Orsay, 20 in Lille and 20 in Sophia).   190

6.4   Execution times of the multisplitting method coupled to the MUMPS solver for different sizes of generated matrices with 120 machines located in 4 sites (40 in Rennes, 40 in Orsay, 25 in Nancy and 15 in Lille).   191

6.5   Execution times of the multisplitting method coupled to the MUMPS or SuperLU solvers for different sizes of generated matrices with 190 machines located in 5 sites (30 in Rennes, 30 in Sophia, 70 in Orsay, 30 in Lyon and 30 in Lille).   192

6.6   Execution times of the multisplitting method coupled to the SparseLib solver for generated square matrices of size 30.10^6 with 200 bi-processors located in 2 sites (120 in Paris, 80 in Nice), so 400 CPUs.   193

6.7   Impacts of memory requirements of the synchronous multisplitting method with SuperLU for the *cage*12 matrix.   195

6.8   Execution times of the multisplitting-Newton method coupled to the MUMPS solver for different sizes of the advection-diffusion problem with 120 machines located in 4 sites and a discretization time step of 360 s.   198

6.9   Execution times of the multisplitting-Newton method coupled to the MUMPS solver for different sizes of the advection-diffusion problem with 120 machines located in 4 sites and a discretization time step of 720 s.   198

6.10 Ratios between synchronous and asynchronous execution times of the multisplitting-Newton method for different sizes and discretization time steps of the advection-diffusion problem with 120 machines located in 4 sites.   199

List of Figures

2.1 Splitting of the matrix. 15
2.2 Spectral radius of the iteration matrices. 23
2.3 Illustration of the Newton method. 42

3.1 Correspondence between radius-based and Flynn's classification of parallel systems. 53
3.2 General architecture of a parallel machine with shared memory. 54
3.3 General architecture of a parallel machine with distributed memory. 55
3.4 General architecture of a local cluster. 56
3.5 General architecture of a distributed cluster. 58
3.6 Hierarchical parallel systems, mixing shared and distributed memory. 60
3.7 Execution flow of the SISC scheme with two processors. . . . 62
3.8 Execution flow of the SIAC scheme with two processors. . . . 64
3.9 Execution flow of the basic AIAC scheme with two processors. 65
3.10 Execution flow of the sender-side semi-flexible AIAC scheme with two processors. 67
3.11 Execution flow of the receiver-side semi-flexible AIAC scheme with two processors. 67
3.12 Execution flow of the flexible AIAC scheme with two processors. 68

4.1 A splitting of matrix A. 76
4.2 A splitting of matrix A using subset J_l of $l \in \{1, ..., L\}$. . . . 77
4.3 Splitting of the matrix for the synchronous Jacobi method. . 87
4.4 An example with three weighting matrices. 91
4.5 An example of possible splittings with three processors. . . . 92
4.6 Decomposition of the matrix. 93
4.7 An example of decomposition of a 9×9 matrix with three processors and one component overlapped at each boundary on each processor. 95
4.8 Overlapping strategy that uses values computed locally. . . . 97
4.9 Overlapping strategy that uses values computed by close neighbors. 98

4.10 Overlapping strategy that mixes overlapped components with close neighbors. 99

4.11 Overlapping strategy that mixes all overlapped values. 100

4.12 Decomposition of the Newton-multisplitting. 102

4.13 Monotonous residual decreases toward the stabilization according to the contraction norm. 105

4.14 A monotonous error evolution and its corresponding non-monotonous residual evolution. 106

5.1 Iterations of the Newton-multisplitting method. 142

5.2 Decomposition of the multisplitting-Newton. 144

5.3 Iterations of the multisplitting-Newton method. 144

5.4 Decentralized global convergence detection based on the leader election protocol. 148

5.5 Simultaneous detection on two neighboring nodes. 148

5.6 Verification mechanism of the global convergence. 159

5.7 Distinction of the successive phases during the iterative process. 160

5.8 Mechanism ensuring that all the nodes are in representative stabilization at least at the time of global convergence detection. 161

5.9 State transitions in the global convergence detection process. 162

6.1 JACE daemon architecture. 177

6.2 A binomial tree broadcast procedure with 2^3 elements. 180

6.3 An example of VDM. 182

6.4 An example illustrating that some messages are ignored. . . . 184

6.5 The GRID'5000 platform in France. 187

6.6 Example of a generated square matrix. 188

6.7 Impacts of the overlapping for a generated square matrix of size 100000. 194

Acknowledgments

The authors wish to thank the following persons for their useful help during the writing of this book: A. Borel, J-C. Charr, I. Ledy, M. Salomon and P. Vuillemin.

Introduction

Computer science is quite a young research area. However, it has already been subject to several major advance steps which, in general, have been closely linked to the technological progresses of the machine components. It can easily be assumed that the current evolution takes place at the level of the communication networks whose quality, either on the reliability or the efficiency levels, begins to be satisfying on large scales.

Beyond the practical interest of the data transfers, this implies a new vision of the computer tool in scientific computing. Indeed, after the successive eras of the single workstations, of the parallel machines and finally of the local clusters, the last advances in large scale networks have permitted the emergence of *clusters of clusters*. That new concept of *meta-cluster* is defined by a set of computational units (workstations, parallel machines or clusters) scattered on geographically distinct sites. Those meta-clusters are then commonly composed of heterogeneous machines linked together by a communication network generally not complete and whose links are also heterogeneous.

As for parallelism in general, the obvious interest of such meta-clusters is to gather a greater number of machines allowing faster treatments and/or the treatment of larger problems. In fact, the addition of a machine in an existing parallel system, even if that machine is less efficient than the ones already in the system, increases the potential degree of parallelism of that system and thus enhances its performance. Moreover, such an addition also increases the global memory capacity of the system which thus allows the storage of more data and then the treatment of larger problems. So, the heterogeneity of the machines does not represent any particular limitation in meta-clusters. Besides, its management has already been intensively studied in the context of local clusters. Nevertheless, a new problem arises with meta-clusters and consists in the efficient management of the heterogeneous communication links. That point is still quite unexplored.

However, it must be noticed that each hardware evolution often comes with a software evolution. Indeed, it is generally necessary to modify or extend the programming paradigms to fully exploit the new capabilities of the machines, the obvious goal always being a gain either in the quality of the results or in the time to obtain them, and if possible, in both of them. Hence, in the same way the parallel machines and local clusters have induced the developments of communication libraries in the programming languages, the emergence of meta-clusters implies an updating of the parallel programming schemes to take into account the specificities of those new computational systems.

In that particular field of parallel programming, the commonly used model is the synchronous message passing. If that model is completely satisfying with parallel machines and local clusters, it is no more the case with meta-clusters. In fact, even if distant communications are faster and faster, they are still far slower than the local ones. So, using synchronous communications in the programming of a meta-cluster is strongly penalizing due to the distant communications between the different sites.

Hence, it seems essential to modify that model or use another model to efficiently use the meta-clusters. Yet, there exists another communication mode between machines which allows, at least partially, to overcome those communication constraints, *the asynchronism*. The principle of that communication scheme is that it does not block the progress of the algorithm. So, during a communication, the sender does not wait for the reception of the message on the destination. Symmetrically, there is no explicit waiting period for message receptions on the receiver and the messages are managed as soon as they arrive. That allows the performance of an implicit and efficient overlapping of the communications by the computations.

Unfortunately, that kind of communication scheme is not usable in all the types of algorithms. However, it is fully adapted to iterative computations. Contrary to the direct methods, which give the solution of a problem in a fixed number of computations, the iterative algorithms proceed by successive enhancements of the approximation of the solution by repeating a same computational process an unknown number of times. When the successive approximations actually come closer to the solution, it is said that the iterative process *converges*.

In the parallel context, those algorithms present the major advantage of allowing far more flexible communication schemes, especially the asynchronous one. In fact, under some conditions which are not too restrictive, the data dependencies between the different computational nodes are no more strictly necessary at each solicitation. In this way, they act more as a progression factor of the iterative process. Moreover, numerous scientific problems can be solved by using that kind of algorithm, especially in PDE (partial differential equations) and ODE (ordinary differential equations) problems. There are even some nonlinear problems, like the polynomial roots problem, which can only be solved by iterative algorithms. Finally, in some other cases such as linear problems, those methods require less memory than the direct ones. Thus, the interest of those algorithms is quite obvious in parallel computations, especially when used on meta-clusters with asynchronous communications.

The objective of this book is to provide theoretical and practical knowledge in parallel numerical algorithms, especially in the context of grid computing and with the specificity of asynchronism. It is written in a way that makes it useful to non-specialists who would like to familiarize themselves with the domain of grid computing and/or numerical computation as well as to researchers specifically working on those subjects. The chapters are organized in progressive levels of complexity and detail. Inside the chapters, the presentation is also progressive and generally follows the same organization: a theoretical part in which the concepts are presented and discussed, an algorithmic part where examples of implementations or specific algorithms are fully detailed, and a discussion/evaluation part in which the advantages and drawbacks of the algorithms are analyzed. The pedagogical aspect has not been neglected and some exercises are proposed at the end of the parts in which this is relevant.

The overall organization of the book is as follows. The first two chapters introduce the general notions on sequential iterative algorithms and their applications to numerical problems. Those bases, required for the following of the book, are particularly intended for students or researchers who are new to the domain. These two chapters recall the basic and essential convergence results on iterative algorithms. First, we consider linear systems and we recall the basic linear iterative algorithms such as Jacobi, Gauss-Seidel and overrelaxation algorithms and then we review iterative algorithms based on minimization techniques such as the conjugate gradient and GMRES algorithms. Second, we consider the Newton method for the solution of nonlinear problems.

Then, the different kinds of parallel systems and parallel iterative algorithms are described in Chapter 3. That chapter also points out the best adequacies of parallel systems and parallel iterative algorithms.

In Chapter 4, parallel synchronous iterative algorithms for numerical computation are provided. Both linear and nonlinear cases are treated and the specific aspects of those algorithms, such as the convergence detection or their implementation, are addressed. In this chapter, we are interested in so-called *multisplitting algorithms*. These algorithms include the discrete analogues of Schwarz multi-subdomains methods and hence are very suitable for distributed computing on distant heterogeneous clusters. They are particularly well suited for physical and natural problems modeled by elliptic systems and discretized by finite difference methods with natural ordering. The parallel versions of minimization such as the methods exposed in Chapter 2 are not detailed in Chapter 4 but it should be mentioned that, thanks to the multisplitting approach, these methods can be used as *inner iterations* of *two-stage* multisplitting algorithms.

The pendant versions of the algorithms introduced in Chapter 4 are proposed in the asynchronous context in Chapter 5. In that part, in addition to the points similarly addressed in the previous chapter, the advantages of asynchronism are pointed out followed by the mathematical model and the representative convergence situations which include *M-matrices and H-matrices*. The multisplitting approach makes it possible to carry out with coarse grained parallelism and to ensure the convergence of their asynchronous versions for a wide class of scientific problems. They are thus very adequate in a context of grid computing, when the ratio *computation time/communication time* is weak. This is why we chose to devote Chapter 4 and Chapter 5 to them. Those last two chapters are particularly aimed at graduate students and researchers.

Finally, Chapter 6 is devoted to the programming environments and experimental results. In particular, the features required for an efficient implementation of asynchronous iterative algorithms are given. Also, numerous experiments led in different computational contexts for the two kinds of numerical problems, linear and nonlinear, are presented and analyzed.

In order to facilitate the reading of the book, the mathematical results which are useful in some chapters but which do not represent the central points of the addressed subjects have been placed in an appendix.

Chapter 1

Iterative Algorithms

1.1 Basic theory

This first chapter introduces the general notions on sequential iterative algorithms.

1.1.1 Characteristic elements of a matrix

Throughout this book, $\mathbb{R}^n$ will denote the real n-dimensional linear space of column vectors x whose components are real numbers $x_1, x_2, ..., x_n \in \mathbb{R}$, the real linear space of dimension 1. The complex n-dimensional linear space $\mathbb{C}^n$ is defined in the same way, with complex numbers $x_i \in \mathbb{C}$, the complex linear space of dimension 1.

The transpose vector of a column vector x is the row vector x^T defined by

$$
x = \begin{pmatrix} x_1 \\ x_2 \\ \vdots \\ x_n \end{pmatrix} \Leftrightarrow x^T = \begin{pmatrix} x_1 \ x_2 \ \cdots \ x_n \end{pmatrix}.
$$

The conjugate transpose of a complex vector is defined in the same way and will be denoted by x^H.

A real (respectively complex) $m \times n$ matrix $A = (A_{i,j})$ defines a linear mapping from $\mathbb{R}^n$ (respectively $\mathbb{C}^n$) to $\mathbb{R}^m$ (respectively $\mathbb{C}^m$). For simplicity's sake we shall write $A \in \mathbb{R}^{m \times n}$ (respectively $\mathbb{C}^{m \times n}$).

The transpose of a square matrix A will be denoted by A^T; it is defined by

$$
\left(A^T \right)_{i,j} = (A_{j,i}).
$$

The conjugate transpose of a complex matrix A is the matrix whose elements are the conjugate elements of the transpose of A. It is denoted by A^H.

A square real (respectively complex) matrix A is symmetric (respectively Hermitian) if $A = A^T$ (respectively, $A = A^H$).

A real matrix A is invertible (or nonsingular) if the linear operator it defines is one-to-one. The inverse of A will be denoted A^{-1}.

For an $n \times n$ matrix A, a scalar λ is called an *eigenvalue* of A if the equation

$$Ax = \lambda x$$

has a non-zero solution. The non-zero vector x is then called an *eigenvector* of A associated to the eigenvalue λ.

The eigenvalues of an $n \times n$ matrix A are the roots of the characteristic polynomial $\det(A - \lambda I)$ where I denotes the $n \times n$ identity matrix

$$I = diag(1, 1, ..., 1) = \begin{pmatrix} 1 & 0 & \cdots & 0 \\ 0 & \ddots & & \vdots \\ \vdots & & \ddots & \\ \vdots & & ... & 0 & 1 \end{pmatrix}.$$

So, λ is an eigenvalue of $A \Leftrightarrow \det(A - \lambda I) = 0$.

DEFINITION 1.1 *Let A be a square real matrix with eigenvalues $\lambda_1, ..., \lambda_n$, then the real number $\rho(A) = \max_{1 \leq i \leq n} |\lambda_i|$ is called the spectral radius of A.*

In this book $T^{(k)}$ will denote either a linear or a nonlinear operator (matrix or function) depending on the iteration k, while T^k will denote the kth power of T.

1.1.2 Norms

Below, we recall basic definitions and results on vectorial norms.

DEFINITION 1.2 *A vectorial norm $\| \, . \, \|$ is a mapping from $\mathbb{R}^n$ ($\mathbb{C}^n$) to $\mathbb{R}$ ($\mathbb{C}$) which satisfies the following three conditions:*

(a) $\|x\| \geq 0$, $\forall x \in \mathbb{R}^n$ ($\mathbb{C}^n$) *and* $\|x\| = 0 \Leftrightarrow x = 0$,

(b) $\|\alpha x\| = |\alpha| \, \|x\|$, $\forall x \in \mathbb{R}^n$ ($\mathbb{C}^n$), $\forall \alpha \in \mathbb{R}$ ($\mathbb{C}$),

(c) $\|x + y\| \leq \|x\| + \|y\|$, $\forall x, y \in \mathbb{R}^n$ ($\mathbb{C}^n$).

Example 1.1
For $p \in [1, +\infty[$, the l_p norms are defined on $\mathbb{R}^n$ or $\mathbb{C}^n$ by

$$\|x\|_p = \left(\sum_{i=1}^{n} |x_i|^p \right)^{1/p}.$$

The Euclidean norm corresponds to the particular case $p = 2$,

$$\|x\|_2 = \sqrt{\sum_{i=1}^{n} |x_i|^2}.$$

The limit case is the l_∞ norm, also called the *maximum norm*,

$$\|x\|_p = \max_{1 \le i \le n} |x_i|.$$

$\square$

DEFINITION 1.3   *We say that a sequence of vectors $x^{(k)}$ converges to a vector x^* if each component $x_i^{(k)}$ converges to x_i^*.*

$$\lim_{k \to +\infty} x^{(k)} = x^* \Leftrightarrow \forall i \in \{1, ..., n\}, \quad \lim_{k \to +\infty} x_i^{(k)} = x_i^*.$$

It is proved that

$$\lim_{k \to +\infty} x^{(k)} = x^* \Leftrightarrow \lim_{k \to +\infty} \left\| x^{(k)} - x^* \right\| = 0,$$

for any arbitrary norm.

We recall below the notion of norms of matrices.

DEFINITION 1.4   *Given any two norms $\| \, . \, \|$ and $\|.\|^1$ respectively on $\mathbb{R}^n$ and $\mathbb{R}^m$, and an $m \times n$ matrix A, the norm of the matrix A with respect to $\| \, . \, \|$ and $\| \, . \, \|^1$ is defined by*

$$\|A\| = \sup_{\|x\|=1} \|Ax\|^1.$$

A matrix norm as defined above satisfies the following properties:

(a) $\|A\| \ge 0, \forall A \in \mathbb{R}^{m \times n}$ and $\|A\| = 0 \Leftrightarrow A = 0$,

(b) $\|\alpha A\| = |\alpha| \, \|A\|, \forall A \in \mathbb{R}^{m \times n}, \forall \alpha \in \mathbb{R}$,

(c) $\|A + B\| \le \|A\| + \|B\|, \forall A, B \in \mathbb{R}^{m \times n}$.

In the particular and important case where $n = m$ and $\| \, . \, \| = \| \, . \, \|^1$, we have the additional property

(d) $\|AB\| \le \|A\| \, \|B\|, \forall A, B \in \mathbb{R}^{n \times n}$.

Example 1.2
Let A be an $m \times n$ matrix, then for $i = 1, 2$ and ∞, the l_i norms are given by

$$\|A\|_1 = \max_{1 \le j \le n} \sum_{i=1}^{m} |A_{i,j}| \tag{1.1}$$

$$\|A\|_\infty = \max_{1 \le i \le m} \sum_{j=1}^{n} |A_{i,j}| \qquad (1.2)$$

$$\|A\|_2 = \sqrt{\lambda}, \qquad (1.3)$$

where λ is the maximum eigenvalue of $A^T A$.

Note that for a symmetric real square matrix with eigenvalues $\lambda_1, ..., \lambda_n$

$$\|A\|_2 = \max_{1 \le i \le n} |\lambda_i| = \rho(A).$$

The Frobenius norm is defined by

$$\|A\|_F^2 = \sum_{i=1}^{m} \sum_{j=1}^{n} |A_{i,j}|^2 \ .$$

⬜

DEFINITION 1.5   *We say that a matrix norm $\|A\|$ is compatible with a vectorial norm $\|x\|$ if*

$$\|Ax\| \le \|A\| \, \|x\| \,.$$

The norms $\|A\|_1$, $\|A\|_\infty$ and $\|A\|_F^2$ are respectively compatible with the norms $\|x\|_1$, $\|x\|_\infty$ and $\|x\|_2$.

DEFINITION 1.6   *We say that a sequence of matrices $A^{(k)}$ converges to a matrix A^* if for all $i, j \in \{1, ..., m\} \times \{1, ..., n\}$, the component $A_{i,j}^{(k)}$ converges to $A_{i,j}^*$.*

Like in the vectorial case, it is shown that

$$\lim_{k \to +\infty} A^{(k)} = A^* \Leftrightarrow \lim_{k \to +\infty} \left\| A^{(k)} - A^* \right\| = 0,$$

for any arbitrary matrix norm.

The following useful results are proved for example in [113] and [93].

PROPOSITION 1.1
For any matrix norm $\| \, . \, \|$, we have

$$\rho(A) \le \|A\| \,.$$

PROPOSITION 1.2
Let A be an $n \times n$ matrix. Then, given any $\varepsilon > 0$, there exists a matrix norm $\| \, . \, \|$ such that

$$\|A\| \le \rho(A) + \varepsilon.$$

The following theorem is fundamental for the study of iterative algorithms.

THEOREM 1.1
Let A be a square matrix, then the following four conditions are equivalent:

1. $\lim_{k \to +\infty} A^k = 0$,

2. $\lim_{k \to +\infty} A^k x = 0$, *for any vector x,*

3. *the spectral radius $\rho(A)$ satisfies: $0 \leq \rho(A) < 1$,*

4. *there exists a matrix norm $\| \, . \, \|$ such that $\|A\| < 1$.*

PROOF 1) $\Rightarrow$ 2) : Let $\| \, . \, \|$ be any matrix norm compatible with a vectorial norm $\| \, . \, \|$. Then we have for any vector x

$$\left\| A^k x \right\| \leq \left\| A^k \right\| \, \|x\| \, ,$$

so $\lim_{k \to +\infty} A^k = 0 \implies \lim_{k \to +\infty} A^k x = 0$.
2) $\Rightarrow$ 3) : Suppose that $\rho(A) \geq 1$, then there exists λ with $|\lambda| \geq 1$ and $x \neq 0$ such that $Ax = \lambda x$. This implies that $A^k x = \lambda^k x$ and then $\lim_{k \to +\infty} A^k x \neq 0$, which contradicts 2).
3) $\Rightarrow$ 4) : Proposition 1.2 says that for a sufficiently small $\varepsilon > 0$, there exists a matrix norm such that $\|A\| \leq \rho(A) + \varepsilon < 1$.
4) $\Rightarrow$ 1) : Since $\left\| A^k \right\| \leq \|A\|^k$, we deduce 1) from 4).   ▯

1.2   Sequential iterative algorithms

Let T be a linear or nonlinear mapping from E to E, whose domain of definition is $D(T)$,

$$T : D(T) \subset E \to E.$$

DEFINITION 1.7   *We say that the mapping T is Lipschitz continuous on a set $D \subset D(T)$ if there exists a constant $L \geq 0$, such that*

$$\forall x, y \in D, \ \|T(x) - T(y)\| \leq L \, \|x - y\| \, .$$

If $L \in \,]0, 1[$ then T is called a contraction (or a contractive mapping), L is then its constant of contraction.

Note that the notion of contraction depends on the considered norm, so that a mapping may be contractive with respect to a norm and not contractive with respect to another norm. Note also that if T is a matrix, then by the definition of a norm and by Theorem 1.1, we obtain

T is a contraction with respect to the norm $\| \cdot \| \Leftrightarrow \|T\| < 1$.

Consider a sequential iterative algorithm associated to T, i.e., a sequential algorithm defined by

---

Algorithm 1.1 A sequential iterative algorithm
_____

   Given an arbitrary $x^{(0)} \in D(T)$
   for k = 0,1,... **do**
      $x^{(k+1)} \leftarrow T(x^{(k)})$
   end for

Then we have the following result on the convergence of Algorithm 1.1.

THEOREM 1.2

Suppose that T is a contraction on a closed set $D \subset D(T)$ and that α is its constant of contraction. Suppose also that $T(D) \subset D$, then Algorithm 1.1 produces a convergent sequence $x^{(k)}$ whose limit is the unique fixed point x^* of T, i.e.,

$$\lim_{k \to +\infty} x^{(k)} = x^* \ where \ x^* = T(x^*).$$

Moreover, we have the estimation

$$\left\| x^{(k)} - x^* \right\| \leq \frac{\alpha}{1 - \alpha} \left\| x^{(k)} - x^{(k-1)} \right\| \tag{1.4}$$

This important result gives a sufficient condition for the convergence of sequential iterative algorithms. It also ensures the uniqueness of the fixed point of a mapping T. See, e.g., [93] for more developments.

The estimation (1.4) allows us to have a bound of the error made at the kth iteration to approximate the unknown solution x^* in terms of the last iterates $x^{(k)}$ and $x^{(k-1)}$. It is important to observe the impact of the value of α on this estimation.

The condition that T maps D into itself can be weakened by supposing that the sequence generated by Algorithm 1.1 remains for a special choice of the initial guess $x^{(0)}$ in D. This choice may be made using the following result (see [79]).

THEOREM 1.3
Let T be contractive on a closed ball $B(x^{(0)}, r) = \{x, \; \|x - x^{(0)}\| \leq r\}$, with constant α. If $x^{(0)}$ satisfies

$$\left\| T(x^{(0)}) - x^{(0)} \right\| \leq (1 - \alpha)r,$$

then Algorithm 1.1 produces a convergent sequence $x^{(k)}$ whose limit is the unique fixed point x^* of T on $B(x^{(0)}, r)$.

PROOF   It is sufficient to prove that $T(B(x^{(0)}, r)) \subset B(x^{(0)}, r)$, this is deduced by remarking that for $x \in B(x^{(0)}, r)$,

$$\left\| T(x) - x^{(0)} \right\| \leq \left\| T(x) - T(x^{(0)}) \right\| + \left\| T(x^{(0)}) - x^{(0)} \right\|$$
$$\leq \alpha \left\| x - x^{(0)} \right\| + (1 - \alpha)r \leq r.$$

$\square$

Finally, note that in the computer science framework all the balls are closed since the set of representable numbers in computers is finite and that the results above can be extended to a general metric space.

Convergence conditions for sequential algorithms are theoretically described by the convergence results on successive approximation methods. Various results can be found in the literature, see for example [113], [93], [79], [91], [114], [90], [31]. In [87], a general topological context for successive approximation methods is studied. The authors define the notion of approximate contraction which is a generalization of the notion of contraction and which is useful in the study of perturbed successive approximation methods.

Theorems 1.2 and 1.3 ensure the convergence to a fixed point x^* but do not give any information on its exact value. Practically, the iterations produced by Algorithm 1.1 are stopped when a *distance* between two iterates is small enough. Algorithm 1.1 becomes,

---

Algorithm 1.2 A sequential iterative algorithm with a stopping criterion

---

Given an arbitrary $x^{(0)} \in D(T)$ and ε a small positive scalar
repeat
   $x^{(k+1)} \leftarrow T(x^{(k)})$
   $k \leftarrow k + 1$
until $d(x^{(k)}, x^{(k-1)}) \leq \varepsilon$

The scalar ε is a small number related to the accuracy desired by the user and the distance d is defined by the norm $\| \; . \; \|$.

In this book we are interested in the solution of numerical problems with iterative algorithms and their implementation on parallel and distributed computers. In the next section we introduce a standard scientific example as a motivation of iterative computing.

1.3   A classical illustration example

Consider the problem of finding a function u of a variable x, satisfying the following differential equation

$$-\frac{d^2u}{dt^2} = f(x), \ x \in [0,1] \tag{1.5}$$

where f is a known function. Suppose also that u satisfies the boundary conditions

$$u(0) = a, \ u(1) = b. \tag{1.6}$$

The problem (1.5) with the conditions (1.6) is a two-point boundary-value problem. It has a unique solution u. This problem is classical, it describes many steady-state physical problems.

Let us make a discretization of the space x using a fixed step size h:

$$h = \frac{1}{n+1}.$$

We will then compute the approximate values of u at the discrete points $h, 2h, ..., nh$. Let $u_1, u_2, ..., u_n$ denote the approximate values of u at the points $h, 2h, ..., nh$ and $u_0 = u(0) = a$ and $u_{n+1} = u(1) = b$.

Let us use the second central difference scheme in order to discretize Equation (1.5),

$$\frac{d^2u}{dt^2} \simeq \frac{u(x+h) - 2u(x) + u(x-h)}{h^2}.$$

The discrete analogue of Equation (1.5) is then

$$-u_{j+1} + 2u_j - u_{j-1} = h^2 f(jh), \ j = 1, ..., n.$$

We obtain a linear system which has the form

$$
\begin{aligned}
2u_1 - u_2 \quad &= h^2 f(h) + a \\
-u_1 + 2u_2 - u_3 \quad &= h^2 f(2h) \\
\vdots \qquad &\quad \vdots \\
-u_{n-2} + 2u_{n-1} - u_n &= h^2 f((n-1)h) \\
-u_{n-1} + 2u_n \quad &= h^2 f(nh) + b.
\end{aligned}
$$

This linear system is equivalent to

$$
\begin{pmatrix}
2 & -1 & 0 & \cdots & 0 \\
-1 & 2 & -1 & \ddots & \vdots \\
0 & -1 & \ddots & \ddots & 0 \\
\vdots & \ddots & \ddots & \ddots & -1 \\
0 & \cdots & 0 & -1 & 2
\end{pmatrix}
\begin{pmatrix}
u_1 \\
u_2 \\
\vdots \\
\vdots \\
u_n
\end{pmatrix}
=
\begin{pmatrix}
h^2 f(h) + a \\
h^2 f(2h) \\
\vdots \\
h^2 f((n-1)h) \\
h^2 f(nh) + b
\end{pmatrix}.
\tag{1.7}
$$

Then the solution of the differential equation (1.5) leads to the solution of the sparse linear system (1.7). Even if the solution of a linear or nonlinear system obtained by the discretization of a scientific problem is studied from the mathematical point of view (existence, uniqueness, convergence), obtaining correct solutions may be hard or impossible due for example to the numerical stiffness of the problem and to round-off errors. To solve (1.7) one can use *direct algorithms* based on the Gaussian elimination method and its enhancements or iterative algorithms in order to approximate this solution by inexpensive (in terms of storage) repetitive computations.

In this book, we are interested in the construction of convergent efficient iterative algorithms in the framework of sequential, parallel synchronous and parallel asynchronous execution modes. The next chapter is dedicated to the basic iterative algorithms for the solution of numerical problems.

Chapter 2

Iterative Algorithms and Applications to Numerical Problems

Introduction

This chapter introduces linear systems and describes basic linear iterative algorithms such as Jacobi, Gauss-Seidel and overrelaxation algorithms. Then it presents iterative algorithms based on minimization techniques such as the Conjugate Gradient and GMRES algorithms. Finally, the Newton method for the solution of nonlinear problems is also introduced.

2.1 Systems of linear equations

2.1.1 Construction and convergence of linear iterative algorithms

Consider a linear system

$$Ax = b, \ x \in \mathbb{R}^n, \tag{2.1}$$

where $A = (A_{i,j})_{1 \leq i,j \leq n}$ is a square and nonsingular matrix and b is a vector of the form $b = (b_1, ..., b_n)^T$. Let A^{-1} be the inverse matrix of A.

The exact solution of (2.1) is $x = A^{-1}b$. This exact solution is often impossible to obtain due to different kinds of errors, such as round-off errors, truncature errors and data perturbation errors. The computation of A^{-1} is more expansive than numerical algorithms for the approximation of the solution of (2.1). Indeed, to compute A^{-1}, we have to solve n linear systems

$$Ax^{(j)} = e^{(j)}, \ j \in \{1, ..., n\},$$

where $e^{(1)} = (1, 0, ..., 0)^T$, $e^{(2)} = (0, 1, 0, ..., 0)^T, ..., e^{(n)} = (0, ..., 0, 1)^T$. The solutions $x^{(j)}$ represent the jth columns of A^{-1}. We can see that the computation of A^{-1} requires the solution of n linear systems such as (2.1)!

To solve (2.1), two classes of algorithms exist: direct algorithms and iterative ones. Direct algorithms lead to the solution after a finite number of elementary operations. The exact solution is theoretically reached if we suppose that there is no round-off error. In direct algorithms, the number of elementary operations can be predicted independently of the precision of the approximate solution.

Iterative algorithms proceed by successive approximations and consist in the construction of a sequence $\left\{x^{(k)}\right\}_{k\in\mathbb{N}}$ the limit of which is the solution of (2.1)

$$\lim_{k\to\infty} x^{(k)} = A^{-1}b.$$

The iterations are stopped when the desired precision is obtained. See Chapter 4 for more developments.

Linear iterative algorithms can be expressed in the form

$$x^{(k+1)} = Tx^{(k)} + c \text{ with a known initial guess } x^{(0)}. \qquad (2.2)$$

Jacobi, Gauss-Seidel, overrelaxation and Richardson algorithms are linear iterative algorithms. If the mapping T does not depend on the current iteration k, then the algorithm is called *stationary*. In the opposite case, the algorithm is called *nonstationary*. The iterations generated by (2.2) correspond to the Picard successive approximations method associated to T. To obtain such algorithms, the fixed point of T has to coincide with the solution of (2.1). For that, the matrix A is partitioned into

$$A = M - N \qquad (2.3)$$

where M is a nonsingular matrix. The linear system (2.1) can thus be written

$$Mx = Nx + b$$

or equivalently to the *fixed point equation*

$$x = M^{-1}Nx + M^{-1}b. \qquad (2.4)$$

From this last equation, the following iterative algorithm is deduced

$$x^{(k+1)} = M^{-1}Nx^{(k)} + M^{-1}b, \text{ with a given } x^{(0)},$$

which has the form of the linear iterative algorithm (Algorithm 2.2).

DEFINITION 2.1   *The linear iterative algorithm (Algorithm 2.2) is convergent to the solution of the linear system (2.1) if given any $x^{(0)} \in \mathbb{R}^n$, $\lim_{k\to+\infty} x^{(k)} = A^{-1}b$.*

The following theorem whose proof is a deduction of Theorem 1.1 of Chapter 1 is essential for the study of the convergence of iterative algorithms; see, e.g., [113].

THEOREM 2.1
Consider a linear system $Ax = b$ whose solution is x^*, then the following three conditions are equivalent:

1. *the linear iterative algorithm (Algorithm 2.2) is convergent,*

2. $\rho(T) < 1$,

3. *there exists a matrix norm $\| \, . \, \|$ such that $\|T\| < 1$.*

Therefore, to build a convergent linear iterative algorithm in order to solve a linear system $Ax = b$, the splitting (2.3) has to satisfy one of the last two conditions of the above theorem, where T is replaced by $M^{-1}N$.

2.1.2 Speed of convergence of linear iterative algorithms

In the above section we have explained how to build a convergent linear iterative algorithm; the convergence is ensured if the spectral radius of the iteration matrix T is strictly less than one, i.e., if the iteration matrix is a contraction. This result is a particular case of the general convergence result given in Chapter 1. The goal of this section is to give tools and results to evaluate the speed of convergence of an iterative algorithm and then to compare iterative linear methods. For more details, the reader is invited to see [113].

LEMMA 2.1
Consider a square iterative matrix T, then for any matrix norm $\|.\|$ we have,

$$\rho(T) = \lim_{k \mapsto +\infty} \left(\left\| T^k \right\| \right)^{1/k}$$

PROOF Proposition 1.1 ensures that

$$(\rho(T))^k = \rho(T^k) \leq \left\| T^k \right\|,$$

so,

$$\rho(T) \leq \left(\left\| T^k \right\| \right)^{1/k}.$$

Consider for an arbitrary ε the matrix $T^{(\varepsilon)} = \frac{T}{\rho(T)+\varepsilon}$, then $\rho(T^{(\varepsilon)}) < 1$ and $\lim_{k \mapsto +\infty} \left(\left\| \left(T^{(\varepsilon)} \right)^k \right\| \right) = 0$. This implies that there exists $K \in \mathbb{N}$ such that $\forall k \geq K, \left\| \left(T^{(\varepsilon)} \right)^k \right\| < 1$. This is equivalent to

$$\forall k \geq K, \ \left(\left\| T^k \right\| \right)^{1/k} < \rho(T) + \varepsilon. \tag{2.5}$$

Since ε is arbitrary, we deduce the lemma. □

Consider a convergent linear iterative algorithm whose iteration matrix T is convergent, i.e., $\rho(T) < 1$. Thus $\lim_{k \to +\infty} x^{(k)} = x^*$. Let us denote by $\varepsilon^{(k)}$ the error vector at iteration k,

$$\varepsilon^{(k)} = x^{(k)} - x^*,$$

then we have

$$\varepsilon^{(k)} = T\varepsilon^{(k-1)} = T^k \varepsilon^{(0)}.$$

Let us choose ε such that $\rho(T) + \varepsilon < 1$. Then the above equality and (2.5) both give

$$\forall k \geq K, \ \left\| \varepsilon^{(k)} \right\| \leq (\rho(T) + \varepsilon)^k \left\| \varepsilon^{(0)} \right\|.$$

So, the speed of convergence of a linear iterative algorithm with iteration matrix T is determined by the spectral radius of T. The smaller the spectral radius is, the faster the algorithm is.

We see that the behavior of the iterative algorithm (Algorithm 2.2) is completely determined by the *fixed point mapping* defined by the fixed point equation (2.4)

$$x \mapsto M^{-1}Nx + M^{-1}b,$$

this is also true in the case of nonlinear systems, as we will see in Section 2.2. So, we will talk about an iterative algorithm associated to a fixed point mapping.

The following definition [113] gives the average rate of convergence and allows the comparison of two iterative algorithms.

DEFINITION 2.2   *Let A be an $n \times n$ matrix. If for some integer m, $\|A^m\| < 1$, then*

$$R(A^m) = -\ln \left[\| A^m \|^{1/m} \right] = -\frac{\ln \|A^m\|}{m}$$

is the average rate of convergence for m iterations of the matrix A. Consider two convergent linear iterative algorithms (I) and (II) with respective iteration matrices A_1 and A_2. If

$$R(A_1^m) > R(A_2^m)$$

then the Algorithm (I) is faster for m iterations than Algorithm (II). The asymptotic rate of convergence of an iterative method with iteration matrix A is defined by

$$R_\infty(A) = \lim_{m \to +\infty} R(A^m) = -\ln \rho(A).$$

It should be noticed that for all m such that $\|A^m\| < 1$ we have $R_\infty(A) \geq R(A^m)$.

In the next sections, we particularize the previous results to basic linear algorithms. We show how to build them and we outline their convergence conditions.

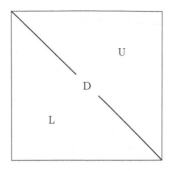

FIGURE 2.1: Splitting of the matrix.

2.1.3   Jacobi algorithm

The Jacobi method is the simplest method to solve a linear system. It consists in decomposing the matrix A into $A = M - N$ where $M = D$ is the diagonal matrix of A. The diagonal elements are assumed to be non-null. An example of the decomposition is given in Equation (2.6).

$$
A = \begin{pmatrix} 10 & -4 & -2 & 0 \\ -2 & 23 & -5 & -1 \\ -1 & -2 & 16 & -7 \\ 0 & -1 & -4 & 18 \end{pmatrix}, \; M = D = \begin{pmatrix} 10 & 0 & 0 & 0 \\ 0 & 23 & 0 & 0 \\ 0 & 0 & 16 & 0 \\ 0 & 0 & 0 & 18 \end{pmatrix}, \; N = \begin{pmatrix} 0 & 4 & 2 & 0 \\ 2 & 0 & 5 & 1 \\ 1 & 2 & 0 & 7 \\ 0 & 1 & 4 & 0 \end{pmatrix}
\tag{2.6}
$$

Using the decomposition of Figure 2.1 which can summarize the Jacobi algorithm and the Gauss-Seidel one (see the next subsection), $M = D$ and $N = -(L + U)$. With that decomposition and Equation (2.1) and following the construction method of the previous section, we obtain

$$
Dx^{(k+1)} = Nx^{(k)} + b.
\tag{2.7}
$$

As D is a diagonal matrix with non-zero elements, the inverse D^{-1} of D contains the inverse of each element of the diagonal and Equation (2.7) gives

$$
x^{(k+1)} = -D^{-1}(L + U)x^{(k)} + D^{-1}b.
\tag{2.8}
$$

It should be noticed that at each iteration, each component of the vector $x^{(k+1)}$ uses components of the previous iteration $x^{(k)}$, so we have to store all the components of $x^{(k)}$ in order to compute $x^{(k+1)}$.

The component-wise form of the Jacobi method is:

$$
x_i^{(k+1)} = (b_i - \sum_{j \neq i} A_{i,j} x_j^{(k)})/A_{i,i}.
\tag{2.9}
$$

In order to implement the Jacobi algorithm, several equivalent variants are possible, depending on whether the values of the matrix may be changed or not and depending on the storage mode of the matrix. Of course this remark is true for almost all the numerical algorithms. Considering that we have the initial matrix A, either the algorithm divides each value of a line by the diagonal element at each iteration, or this transformation is performed before the Jacobi algorithm.

Algorithm 2.1 presents a possible implementation of the Jacobi method. In the following algorithm we consider that A is a two dimensional array that contains the elements of the matrix. Each dimension of A has $Size$ elements. We consider that a structure (vector or matrix) of $Size$ elements is numbered from 0 to $Size - 1$ as this is traditionally the case in the C programming language. The unknown vector at the iteration $k + 1$, $x^{(k+1)}$, is stored into X, a one dimensional vector of size $Size$. In the same way, $x^{(k)}$ is represented by the one dimensional array $XOld$.

The principle of this algorithm consists in iterating on the following statements until the stopping criterion is reached. Each element $X[i]$ contains the product of the line i of the matrix A multiplied by the previous vector ($XOld$) except for element i. The purpose of the last step of an iteration is to take into account the right-hand side and to divide all the results by the diagonal element i.

Algorithm 2.1 Jacobi algorithm

Size: size of the matrix
X[Size]: solution vector
XOld[Size]: solution vector at the previous iteration
A[Size][Size]: matrix
B[Size]: right-hand side vector

repeat
  for i=0 to Size−1 **do**
    X[i] ← 0
    for j=0 to i−1 **do**
      X[i] ← X[i]+A[i][j]×XOld[j]
    end for
    for j=i+1 to Size−1 **do**
      X[i] ← X[i]+A[i][j]×XOld[j]
    end for
  end for
  for i=0 to Size−1 **do**
    XOld[i] ← (B[i]−X[i])/A[i][i]
  end for
until stopping criteria is reached

2.1.4   Gauss-Seidel algorithm

The Gauss-Seidel method presents some similarities with the Jacobi method. The decomposition is slightly different since N is decomposed into two parts: a strict lower part L and a strict upper part U as in Figure 2.1.

In opposition to the Jacobi method which only uses components of the previous iteration to compute the current one, the Gauss-Seidel method uses all the components that have already been computed during the current iteration to compute the other ones. The Gauss-Seidel method is defined by

$$Dx^{(k+1)} + Lx^{(k+1)} + Ux^{(k)} = b. \tag{2.10}$$

The components that have been computed at the current iteration are represented by the lower part L. Equation (2.10) can be rewritten as

$$x^{(k+1)} = -(D+L)^{-1}Ux^{(k)} + (D+L)^{-1}b.$$

The component-wise form of the Gauss-Seidel method is:

$$x_i^{(k+1)} = (b_i - \sum_{j<i} A_{i,j}x_j^{(k+1)} - \sum_{j>i} A_{i,j}x_j^{(k)})/A_{i,i}$$

As mentioned in the previous section both the Jacobi method and the Gauss-Seidel method can be written as

$$x^{(k+1)} = M^{-1}Nx^{(k)} + M^{-1}b$$

in which $A = M - N$ is a splitting of A, $M = D$, $N = -(L+U)$ for Jacobi and $M = D + L$, $N = -U$ for Gauss-Seidel. Then the iteration matrix of the Jacobi algorithm is

$$J = -D^{-1}(L+U)$$

and the iteration matrix of the Gauss-Seidel algorithm is

$$\mathcal{L}_1 = -(D+L)^{-1}U.$$

Using the previous notations for the Jacobi algorithm, it is possible to write the Gauss-Seidel Algorithm 2.2. The principle of this algorithm is very similar to the Jacobi one. As elements before i use the current iteration vector and the elements after i use the previous iteration vector, it is necessary to use an intermediate variable to store the result. In this algorithm, we use the variable V. Apart from that difference, the rest of the algorithm is similar to the Jacobi one.

The Stein-Rosenberg theorem [108] is based on the Perron-Frobenius theory on nonnegative matrices (see the Appendix) [97], [58]. It allows the comparison of the asymptotic rates of convergence of the point Jacobi and the Gauss-Seidel methods. Its proof can be found in [113].

Algorithm 2.2 Gauss-Seidel algorithm

Size: size of the matrix
X[Size]: solution vector
XOld[Size]: solution vector at the previous iteration
A[Size][Size]: matrix
B[Size]: right-hand side vector
V: intermediate variable

repeat
  for i=0 to Size−1 **do**
    V ← 0
    for j=0 to i−1 **do**
      V ← V+A[i][j]×X[j]
    end for
    for j=i+1 to Size−1 **do**
      V ← V+A[i][j]×XOld[j]
    end for
    X[i] ← (B[i]−V)/A[i][i]
  end for
  for i=0 to Size−1 **do**
    XOld[i] ← X[i]
  end for
until stopping criteria is reached

THEOREM 2.2

Consider a linear system $Ax = b$ where $A = L + D + U$. Suppose that the Jacobi matrix $J = -D^{-1}(L + U)$ is nonnegative, then, the spectral radii of the iteration matrices of Jacobi and Gauss-Seidel satisfy one of the following exclusive conditions:

1. $\rho(J) = \rho(\mathcal{L}_1) = 0$,

2. $0 < \rho(\mathcal{L}_1) < \rho(J) < 1$,

3. $1 = \rho(J) = \rho(\mathcal{L}_1)$,

4. $1 < \rho(J) < \rho(\mathcal{L}_1)$.

So, if the Jacobi matrix is nonnegative, the Jacobi and the Gauss-Seidel algorithms are simultaneously convergent or divergent. As a corollary we obtain the following comparison between the asymptotic rates of convergence

$$R_\infty(\mathcal{L}_1) > R_\infty(J).$$

So, the asymptotic rate of convergence of the Gauss-Seidel method is higher than the Jacobi one.

2.1.5   Successive overrelaxation method

The successive overrelaxation method, or SOR, can be obtained by applying extrapolation to the Gauss-Seidel method. It consists in mixing the form of a weighted average between the previous iterate and the computed Gauss-Seidel iterate for each component $x_i^{(k+1)}$,

$$x_i^{(k+1)} = \omega \overline{x}_i^{(k+1)} + (1 - \omega) x_i^{(k)}$$

where $\overline{x}$ represents the Gauss-Seidel iterate, and ω is a relaxation parameter. By choosing an appropriate ω, it is possible to increase the speed of convergence to the solution.

So we obtain

$$x_i^{(k+1)} = (1 - \omega) x_i^{(k)} + \omega (b_i - \sum_{j<i} A_{i,j} x^{(k+1)} - \sum_{j>i} A_{i,j} x^{(k)})/A_{i,i}.$$

In matrix terms, the SOR algorithm can be written as follows:

$$x^{(k+1)} = -(D + \omega L)^{-1}(\omega U - (1 - \omega)D)x^{(k)} + \omega(D + \omega L)^{-1}b,$$

or equivalently

$$\left(\frac{\omega L + D}{\omega}\right) x^{(k+1)} = \left(\frac{(1 - \omega)D - \omega U}{\omega}\right) x^{(k)} + b,$$

so the successive overrelaxation algorithm is a particular linear iterative algorithm corresponding to the decomposition $A = M - N$ where $M = \frac{\omega L + D}{\omega}$ and $N = \frac{(1-\omega)D - \omega U}{\omega}$. The iteration matrix is

$$\mathcal{L}_\omega = (\omega L + D)^{-1} ((1 - \omega)D - \omega U).$$

In Algorithm 2.3 we can remark that the difference between the SOR implementation and the Gauss-Seidel one only concerns the parameter ω which allows us to take into account an intermediate value between the current iteration and the previous one.

The following theorem which is a corollary of a general theorem due to Ostrowski [94] gives the convergence of the overrelaxation algorithm.

THEOREM 2.3
If the matrix A is symmetric (respectively Hermitian), then the successive overrelaxation algorithm converges for $\omega \in \,]0, 2[$.

If $\omega = 1$, the SOR method becomes the Gauss-Seidel method. In [76] Kahan has proved that SOR fails to converge if ω is outside the interval $]0, 2[$. The term overrelaxation should be used when $1 < \omega < 2$; nevertheless, it is used for any value of $0 < \omega < 2$.

Algorithm 2.3 SOR algorithm

Size: size of the matrix
X[Size]: solution vector
XOld[Size]: solution vector at the previous iteration
A[Size][Size]: matrix
B[Size]: right-hand side vector
V: intermediate variable
Omega: parameter of the method

repeat
  for i=0 to Size−1 **do**
    V ← 0
    for j=0 to i−1 **do**
      V ← V+A[i][j]×X[j]
    end for
    for j=i+1 to Size−1 **do**
      V ← V+A[i][j]×XOld[j]
    end for
    V ← (B[i]−V)/A[i][i]
    X[i] ← XOld[i]+Omega×(V−XOld[i])
  end for
  for i=0 to Size−1 **do**
    XOld[i] ← X[i]
  end for
until stopping criteria is reached

Commonly, the computation of the optimal value of ω for the rate of convergence of SOR is not possible in advance. When this is possible, the computation cost of ω is generally expensive. That is why a solution consists in using some heuristics to estimate it. For example, some heuristics are based on the mesh spacing of the discretization of the physical problem [81].

2.1.6   Block versions of the previous algorithms

The three previous algorithms only work component-wise. The particularity of a block version of an existing algorithm consists in taking into account block components rather than simple components. Consequently, the structure of the algorithm is the same but the computation and the implementation are different.

So, matrix A and vectors x and b are partitioned as follows:

$$
A = \begin{pmatrix} A_{11} & A_{12} & A_{13} & \cdots & A_{1n} \\ A_{21} & A_{22} & A_{23} & \cdots & A_{2n} \\ A_{31} & A_{32} & A_{33} & \cdots & A_{3n} \\ \vdots & \vdots & \vdots & \ddots & \vdots \\ A_{n1} & A_{n2} & A_{n3} & \cdots & A_{nn} \end{pmatrix}, \quad x = \begin{pmatrix} X_1 \\ X_2 \\ X_3 \\ \vdots \\ X_n \end{pmatrix}, \quad b = \begin{pmatrix} B_1 \\ B_2 \\ B_3 \\ \vdots \\ B_n \end{pmatrix}, \quad (2.11)
$$

where x and b are partitioned into subvectors in a compatible way with the partitioning of A.

So, it is possible to define a similar splitting as in Figure 2.1 in which D is composed of diagonal blocks as follows:

$$
D = \begin{pmatrix} A_{11} & & & \\ & A_{22} & & \\ & & A_{33} & \\ & & & \ddots \\ & & & & A_{nn} \end{pmatrix}, \quad E = \begin{pmatrix} 0 & & & & \\ -A_{21} & 0 & & & \\ -A_{31} & -A_{32} & 0 & & \\ \vdots & \vdots & \vdots & \ddots & \\ -A_{n1} & -A_{n2} & \cdots & A_{nn-1} & 0 \end{pmatrix},
$$

$$
F = \begin{pmatrix} 0 & -A_{12} & -A_{13} & \cdots & & -A_{1n} \\ & \ddots & \vdots & \vdots & & \vdots \\ & & 0 & & -A_{n-2n-1} & -A_{n-2n} \\ & & & 0 & & -A_{n-1n} \\ & & & & & 0 \end{pmatrix} \qquad (2.12)
$$

Suppose that we have *NbBlock* blocks that have the same size *BlockSize*. Thus, the offset of the ith block is stored into $i \times BlockSize$. Algorithm 2.4 gives a possible implementation of the block Jacobi algorithm. The first step consists in duplicating the right-hand side into an intermediate variable *BTmp*. Then, for each block k, components corresponding to the block of the right-hand side are updated using the previous iteration vector *XOld*. The corresponding linear subsystem needs then to be solved in order to obtain an approximation of the corresponding unknown vector x. The choice of the method to solve the linear system is free. It may be a direct or an iterative method. When an iterative method is used we talk about *two-stage* iterative algorithms.

The advantage of the block Jacobi method is that the number of iterations is often significantly decreased. The drawback of this method is that it requires the resolution of several linear subsystems which is not an easy task. Moreover, the precision of the inner solver has an influence on the number of iterations required for the outer solver to reach the convergence.

Implementing a block version of the Gauss-Seidel and the SOR methods simply requires the use of the last version of components of previous blocks and the previous version of components of the next blocks (as it is the case in the componentwise version). Moreover, the SOR version needs to include a relaxation parameter as in the componentwise version.

Algorithm 2.4 Block Jacobi algorithm

Size: size of the matrix
BlockSize: size of a block
NbBlock: Number of blocks
A[Size][Size]: matrix
B[Size]: right-hand side vector
BTmp[Size]: intermediate right-hand side vector
X[Size]: solution vector
XOld[Size]: solution vector at the previous iteration

repeat
  for i=0 to Size−1 **do**
    BTmp[i] ← B[i]
  end for
  for k=0 to NbBlock−1 **do**
    for i=k×BlockSize to (k+1)×BlockSize−1 **do**
      for j=0 to k×BlockSize−1 **do**
        BTmp[i] ← BTmp[i]−A[i][j]×XOld[i]
      end for
      for j=(k+1)×BlockSize to Size−1 **do**
        BTmp[i] ← BTmp[i]−A[i][j]×XOld[i]
      end for
    end for
    Solve the linear subsystem corresponding to the kth block of A and
    $BTmp$ (A_{kk}, X_k, B_k)
  end for
until stopping criteria is reached

2.1.7 Block tridiagonal matrices

In this section we review the convergence results in the important case of block tridiagonal matrices

A block tridiagonal matrix A is a matrix of the form

$$A = \begin{pmatrix} A_1 & B_1 & 0 & \cdots & & 0 \\ C_1 & A_2 & B_2 & \ddots & & \vdots \\ 0 & C_2 & \ddots & \ddots & & 0 \\ \vdots & \ddots & \ddots & \ddots & & B_{\alpha-1} \\ 0 & \cdots & 0 & C_{\alpha-1} & & A_\alpha \end{pmatrix}$$

Suppose that in the linear system (2.1), the matrix A is block tridiagonal, then we have the following result (see [113]):

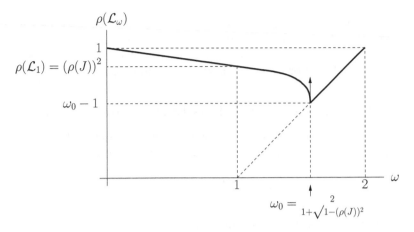

FIGURE 2.2: Spectral radius of the iteration matrices.

THEOREM 2.4
The Jacobi and Gauss-Seidel algorithms converge or diverge simultaneously and

$$\rho(\mathcal{L}_1) = (\rho(J))^2.$$

The following result compares the convergence of Jacobi, Gauss-Seidel and successive overrelaxation algorithms in the case of block tridiagonal matrices.

THEOREM 2.5
Let the matrix A of the linear system (2.1) be block tridiagonal. Suppose that the eigenvalues of the block Jacobi iteration matrix are real. Then the block Jacobi and the block successive overrelaxation algorithms converge or diverge simultaneously. The spectral radius of the iteration matrices varies following Figure 2.2. ω_0 is the optimum parameter corresponding to the lowest spectral radius, its exact value is

$$\omega_0 = \frac{2}{1 + \sqrt{1 - (\rho(J))^2}}$$

and

$$\rho(\mathcal{L}_{\omega_0}) = \omega_0 - 1.$$

An important particular situation where the hypotheses of the above theorem are satisfied is that of symmetric (Hermitian) positive definite matrices.

The next methods are called *nonstationary methods.* Those methods differ from stationary methods because the computations involve information which is led to be changed at each iteration. Usually, constants for nonstationary methods are defined by inner products of residuals or other vectors obtained in the iterative process. In the previous sections we were interested in linear iterative algorithms to solve linear systems of equations. In the next section we will review another class of algorithms to solve linear systems. These algorithms are based on the minimization of a function.

2.1.8   Minimization algorithms to solve linear systems

Assume that we have to solve a linear system of the form (2.1) and that A is symmetric positive definite. Let us consider the function

$$F(x) = \frac{1}{2}(Ax, x) - (b, x) \tag{2.13}$$

where (x, y) denotes the Euclidean scalar product

$$(x, y) = x^T y.$$

Let us denote by $\|x\|_A$ the A-norm of a vector x,

$$\|x\|_A = (Ax, x)^{1/2} \tag{2.14}$$

The derivative of F is

$$\nabla F(x) = F'(x) = Ax - b,$$

and the Hessian of F is

$$\nabla^2 F(x) = A.$$

Since the Hessian of F is symmetric positive definite, the solution of (2.1) coincides with the minimum of F. Thus solving the minimization problem

$$\min F(x), \ x \in \mathbb{R}^n, \tag{2.15}$$

and solving the problem (2.1) are equivalent tasks.

The principle of minimization algorithms is as follows: to solve (2.1) we minimize the function F. To do that we choose an initial guess $x^{(0)}$ and we compute a new iterate on a subspace of $\mathbb{R}^n$ (called a *search* subspace). This subspace is defined by some constraints which define another subspace of $\mathbb{R}^n$ (the *direction* subspace); the aim is to minimize the value of F at each new iterate. We also talk about projection methods in general and orthogonal projection methods when the search subspace and the constraint subspace coincide.

Below we give the principle of descent and gradient algorithms in a one-dimensional projection process and the principles of the Conjugate Gradient, the GMRES and the BiConjugate Gradient algorithms. We simply explain the idea of each algorithm, then we give its main results and its implementation.

2.1.8.1 Descent and Gradient algorithms

The gradient method belongs to the class of numerical methods called *descent methods*. In order to minimize F, we choose an initial point $x^{(0)}$ and we compute a new iterate $x^{(1)}$ such that $F(x^{(1)}) < F(x^{(0)})$. The new iterate $x^{(1)}$ is defined by

$$x^{(1)} = x^{(0)} + p^{(0)} d^{(0)}$$

where $d^{(0)}$ is a non-null vector of $\mathbb{R}^n$ and $p^{(0)}$ is a nonnegative real, so $d^{(0)}$ is chosen so that

$$F(x^{(0)} + p^{(0)} d^{(0)}) < F(x^{(0)})$$

When $d^{(0)}$ exists, it is called a *descent direction* and $p^{(0)}$ is called a *descent step*. Those two values can be constant or changed at each iteration. The general scheme of a descent method is:

$$\begin{cases} x^{(0)} \ given \\ x^{(k+1)} = x^{(k)} + p^{(k)} d^{(k)} \end{cases} \tag{2.16}$$

with $d^{(k)} \in \mathbb{R}^n - \{0\}$ and $p^{(k)} \in \mathbb{R}^{+*}$, and

$$F(x^{(k)} + p^{(k)} d^{(k)}) < F(x^{(k)}) \tag{2.17}$$

A natural idea to find a descent direction consists in making a Taylor development of the function F between two consecutive iterates $x^{(k)}$ and $x^{(k+1)} = x^{(k)} + p^{(k)} d^{(k)}$:

$$F(x^{(k)} + p^{(k)} d^{(k)}) = F(x^{(k)}) + p^{(k)} (\nabla F(x^{(k)}), d^{(k)}) + o(p^{(k)} d^{(k)}) \tag{2.18}$$

In order to have (2.17), it is possible to choose as initial approximation $d^{(k)} = -\nabla F(x^{(k)})$. Hence, we obtain the gradient algorithm

$$\begin{cases} x^{(0)} \ given \\ x^{(k+1)} = x^{(k)} - p^{(k)} \nabla F(x^{(k)}) = x^{(k)} + p^{(k)} (b - Ax^{(k)}) \end{cases} \tag{2.19}$$

From a practical point of view, implementing this algorithm is quite easy. Algorithm 2.5 sums up the gradient method.

Algorithm 2.5 Gradient algorithm

Initializes x and p
repeat
  $x \leftarrow x - p\nabla F(x)$
  compute a new p if needed
until stopping criteria is reached

If we use a variable step at each iteration, we obtain the optimal step gradient method. With this method we choose a step p which minimizes the cost function $\phi(p) = F(x^{(k)}) - p\nabla F(x^{(k)})$ at the iteration k. In the case where F is defined by (2.13) and if we assume that $p^{(k-1)} \neq 0$, we can prove that the optimum value of $p^{(k)}$ is

$$p_{opt} = \frac{\left\| r^{(k-1)} \right\|_2^2}{\left(Ar^{(k-1)}, r^{(k-1)} \right)} \quad \text{where } r^{(k-1)} = b - Ax^{(k-1)}.$$

The quantities $r^{(k)}$ ($k \geq 0$) are called the *residual vectors*; we can see that they have to converge to 0.

In practice, p is defined using a linear search of optimal step using for example the Wolfe's algorithm [119].

The following result gives the convergence behavior of the gradient algorithm; see [72].

THEOREM 2.6

Let A be a symmetric positive definite matrix and let $\lambda_{\min}$ and $\lambda_{\max}$ denote the extreme eigenvalues of A. Then for any initial guess $x^{(0)}$, the gradient algorithm produces iterates which converge to the solution of (2.1). Moreover, F being defined in (2.13), we have the error estimates

$$\left\| x^{(m)} - x^* \right\|_A = \left(\frac{\lambda_{\max} - \lambda_{\min}}{\lambda_{\max} + \lambda_{\min}} \right)^m \left\| x^{(0)} - x^* \right\|_A,$$

and

$$F(x^{(m)}) - F(x^*) = \left(\frac{\lambda_{\max} - \lambda_{\min}}{\lambda_{\max} + \lambda_{\min}} \right)^{2m} \left(F(x^{(0)}) - F(x^*) \right).$$

2.1.8.2   Conjugate gradient algorithm

The conjugate gradient (CG) method results from the combination of the gradient method and the search of optimal directions. Indeed, the gradient method suffers from the fact that even if two successive residual vectors are orthogonal (form a right angle), a residual may not be orthogonal to a non-successive residual vector. This leads to a high number of iterations to find the solution of the linear system. The CG method also belongs to the category of projection methods and it is designed for symmetric positive definite systems. The CG method ensures that any two residual vectors are orthogonal, that property is achieved by a suitable choice of the directions $p^{(k)}$ (which are A-orthogonal in the sense of the scalar product defined in (2.14), i.e., conjugate) as a linear combination of $p^{(k-1)}$ and $r^{(k)}$. The method requires that only a few iterates are kept in memory. At each iteration, two inner products are performed in order to update a scalar which is used to compute the sequence of iterates and residuals satisfying orthogonality conditions. Below we give the different steps of the conjugate gradient algorithm.

At each iteration, the iterates $x^{(k+1)}$ are updated in each direction by a multiple (called $\alpha^{(k+1)}$) of the search direction vector $p^{(k+1)}$

$$x^{(k+1)} = x^k + \alpha^{(k+1)} p^{(k+1)}. \tag{2.20}$$

The residuals are updated in the same way as follows:

$$r^{(k+1)} = r^{(k)} - \alpha^{(k+1)} q^{(k+1)}, \tag{2.21}$$

where $q^{(k+1)} = Ap^{(k+1)}$. The parameter $\alpha^{(k+1)} = (r^{(k)}, r^{(k)})/(Ap^{(k+1)}, p^{(k+1)})$ minimizes $(A^{-1} r^{(k+1)}, r^{(k+1)})$ among all possible choices for α in (2.21). The search directions are updated using the residuals

$$p^{(k+1)} = r^{(k+1)} + \beta^{(k)} p^{(k)}, \tag{2.22}$$

where the parameter $\beta^{(k)} = (r^{(k)}, r^{(k)})/(r^{(k-1)}, r^{(k-1)})$ ensures that vectors $p^{(k+1)}$ and $Ap^{(k)}$ are orthogonal. This is equivalent to the condition that vectors $r^{(k+1)}$ and $r^{(k)}$ are orthogonal.

The residuals build an orthogonal basis for the space $span\{r^0, Ar^0, A^2 r^0 ...\}$.

Algorithm 2.6 illustrates a possible implementation of the CG algorithm. This algorithm uses a matrix notation that can be used with several libraries such as MV++ used in SparseLib [44], or the Matrix Template Library [105].

Let $\mathcal{K}$ denote the condition number $\frac{\lambda_{max}}{\lambda_{min}}$ of the symmetric positive definite matrix A, then it can be proved that [72],

THEOREM 2.7
The CG iterates satisfy the error estimates

$$\| x^{(k)} - x^* \|_A \leq 2 \left(\frac{\sqrt{\mathcal{K}} - 1}{\sqrt{\mathcal{K}} + 1} \right)^k \| x^{(0)} - x^* \|_A .$$

2.1.8.3 GMRES

The GMRES (Generalized Minimum RESidual) method is a projection algorithm in which the constraint subspace is AK_m where K_m is the Krylov subspace $K_m = span \left\{ v^{(1)}, Av^{(1)}, ..., A^{m-1} v^{(1)} \right\}$ and $v^{(1)} = \frac{r^{(0)}}{\|r^{(0)}\|_2}$. This method was designed by Saad and Schultz [103]. It is based on the Arnoldi-modified Gram-Schmidt procedure to build orthogonal basis of the Krylov subspace and it has the property to compute an approximation which minimizes the Euclidean norm of the residual over all vectors of the form $x^{(0)} + K_m$.

To apply the GMRES algorithm, the matrix A has to be positive definite but not necessarily symmetric. The most popular form of the GMRES is based on the modified Gram-Schmidt procedure. In order to control the storage requirements, this method uses restarts. Without them, the GMRES method is ensured to converge in n steps where n is the size of the matrix

Algorithm 2.6 Conjugate Gradient algorithm

Size: size of the matrix
A[Size][Size]: matrix
B[Size]: right-hand side vector
X[Size]: solution vector
R[Size]: residual vector
P[Size]: search direction vector
Q[Size]: orthogonal vector to the search direction
Alpha, Beta, Rho, RhoOld: scalar variables

R ← B−A×X
i ← 1
repeat
   Rho ← (R,R)
   if i=1 **then**
      P ← R
   else
      Beta ← Rho/RhoOld
      P ← R+Beta×P
   end if
   Q ← A×P
   Alpha ← Rho/(P,Q)
   X ← X+Alpha×P
   R ← R−Alpha×Q
   RhoOld ← Rho
   i ← i+1
until stopping criteria is reached

(considering exact arithmetic). However, from a practical point of view, this is inconceivable when n is large because it requires too much memory storage and too many computations. As a consequence, restarts are essential in practice.

The GMRES method generates a sequence of orthogonal vectors which must be stored. As mentioned above, the algorithm uses a modified Gram-Schmidt orthogonalization. The orthogonal basis of the GMRES is formed explicitly by Algorithm 2.7.

Algorithm 2.7 GMRES orthogonal basis construction

$w^{(i)} \leftarrow A \times v^{(i)}$
for k=1 to i **do**
   $w^{(i)} \leftarrow w^{(i)} - (w^{(i)}, v^{(k)})v^{(k)}$
end for
$v^{(i+1)} \leftarrow w^{(i)} / \parallel w^{(i)} \parallel_2$

The GMRES algorithm has the property that the residual norm can be computed without the sequence of the iterates. So, the method works by computing the successive residual norms until they become accurate enough and only then it generates the iterate, which is the time consuming step, according to the following construction:

$$x^{(i)} = x^{(0)} + y_1 v^{(1)} + y_2 v^{(2)} + \cdots + y_i v^{(i)}$$

where the y_k are chosen to minimize the residual norm $\| b - Ax^{(i)} \|$.

As can be seen, the GMRES method requires a large amount of work and storage which rises linearly with the iteration count. This drawback can be overcome by restarting the algorithm after a chosen number m of iterations. In that case, the accumulated data are wiped out and the intermediate results are used in order to compute the next m iterations as initial data.

Unfortunately, the problem is to define an appropriate value for m. Choosing a value which is too low makes the algorithm converge slowly, whereas choosing a value larger than necessary involves excessive work and requires more memory storage.

Algorithm 2.8 illustrates an implementation of the GMRES algorithm. The principle of this algorithm is the following. At each iteration, the algorithm:

1. computes the Arnoldi process which gives the Hessenberg matrix H (using a QR factorization) and the orthogonal basis V,

2. applies the Givens rotations with the functions *ApplyPlaneRotation* and *GeneratePlaneRotation* (by applying the rotations on the previous elements of H, then by computing the new rotations and finally by applying it on the right-hand side vector for the minimization problem),

3. if needed it solves the minimization problems using H and S with the function *Update* in order to update the values of the solution vector X.

The matrix H is used to store the Hessenberg decomposition. It is computed with the Arnoldi method using a QR factorization.

The functions *GeneratePlaneRotation* and *ApplyPlaneRotation* are used to compute the Givens orthogonalization. In these functions, the arrays CS and SN respectively represent the sine and the cosine for the rotations.

The function $UPDATE$ is used to restart the algorithm when the size of the orthogonal basis becomes too large. In this function, the variable K represents the current size of the orthogonal basis. This function aims at finding Y such that it is the solution of $H \times Y = S$. Remember that H is an upper triangular matrix. Then the vector X is updated using the orthogonal basis V and Y.

In [29, 102, 69, 68, 115, 56], interested readers will be able to find more information on the subject. In [102], two interesting results concerning the convergence are described. First a global convergence result is given.

Algorithm 2.8 GMRES algorithm

M: number of iterations between each restart
Size: size of the matrix
A[Size][Size]: matrix
B[Size]: right-hand side vector
H[M+1][m]: matrix for the Hessenberg decomposition
V[M+1][Size]: orthogonal basis for the Krylov subspace
X: solution vector
R: residual vector
S[M+1]: right-hand side vector for the minimization problem
CS[M+1], SN[M+1]: cosine and sine for the Givens rotations
W[Size]: intermediate vector used in the Arnoldi's method to compute H
NormB: norm of B
NormR: norm of R

$R \leftarrow B - A \times X$
$NormB \leftarrow ||B||_2$
$NormR \leftarrow ||R||_2$
repeat
  $V[0] \leftarrow R/NormR$
  $S[0] \leftarrow NormR$
  for i=0 to M−1 **do**
    $W \leftarrow A \times V[i]$
    for k=0 to i **do**
      $H[k][i] \leftarrow (W, V[k])$
      $W \leftarrow W - H[k][i] \times V[k]$
    end for
    $H[i+1][i] \leftarrow ||W||_2$
    $V[i+1] \leftarrow W/H[i+1][i]$
    for k=0 to i−1 **do**
      ApplyPlaneRotation(H[k][i], H[k+1][i], CS[k], SN[k])
    end for
    GeneratePlaneRotation(H[i][i], H[i+1][i], CS[i], SN[i])
    ApplyPlaneRotation(H[i][i], H[i+1][i], CS[i], SN[i])
    ApplyPlaneRotation(S[i], S[i+1], CS[i], SN[i])
    if $|S[i+1]|/NormB$ is small enough **then**
      UPDATE(X, i, M, H, S, V)
      quit
    end if
  end for
  UPDATE(X, M−1, M, H, S, V)
  $R \leftarrow B - A \times X$
  $NormR \leftarrow ||R||_2$
  if NormR/NormB is small enough **then**
    quit
  end if
until stopping criteria is reached
    (NormR $\leq \epsilon$ or a too large number of iterations has been performed)

Algorithm 2.9 Function *ApplyPlaneRotation(DX, DY, CS, SN)*

This function returns DX and DY
Tmp: intermediate variable

Tmp ← CS × DX + SN × DY
DY ← −SN × DX + CS × DY
DX ← Tmp

Algorithm 2.10 Function *GeneratePlaneRotation(DX, DY, CS, SN)*

This function returns CS and SN

if DY=0 **then**
   CS ← 1
   SN ← 0
else
 if $|DY| > |DX|$ **then**
    SN ← $1 \, / \, \sqrt{1 + (DX/DY)^2}$
    CS ← SN×DX/DY
 else
    CS ← $1 \, / \, \sqrt{1 + (DY/DX)^2}$
    SN ← CS×DY/DX
 end if
end if

Algorithm 2.11 Function *UPDATE(X, K, M, H, S, V)*

K: size of the orthogonal basis
Y[K+1]: solution vector of the system $HY = S$

for i=0 to K **do**
   Y[i] ← S[i]
end for
for i=K to 0 **do**
   Y[i] ← Y[i]/H[i][i]
   for j=i−1 to 0 **do**
     Y[j] ← Y[j]−H[j][i]×Y[i]
   end for
end for
for j=0 to K **do**
   X ← X+Y[j]×V[j]
end for

THEOREM 2.8

If A is a positive definite matrix, then the GMRES algorithm converges for any $m \geq 1$ where m is the dimension of the considered Krylov space.

Then a proposition allows the provision of an upper bound on the convergence rate of the GMRES iterates.

PROPOSITION 2.1

Assume that A is a diagonalizable matrix and let $A = X\Lambda X^{-1}$ where $\Lambda = diag\{\lambda_1, \lambda_2, \ldots, \lambda_n\}$ is the diagonal matrix of eigenvalues and X is the matrix of eigenvectors. In the following, $\mathbb{P}_m$ denotes the set of all Chebyshev polynomials of degree m. By defining

$$\epsilon^{(m)} = min_{p \in \mathbb{P}_m, p(0)=1} max_{i=1,\ldots,n} |p(\lambda_i)|, \tag{2.23}$$

it is possible to prove that the residual norm achieved by the mth step of GMRES satisfies the inequality

$$\| r_m \|_2 \leq \mathcal{K}_2(X)\epsilon^{(m)} \| r_0 \|_2 \tag{2.24}$$

where $\mathcal{K}_2(X) \equiv \| X \|_2 \| X^{-1} \|_2$.

Unfortunately, except in the case when X is normal (i.e., $X^H X = X X^H$ so $\| X \|_2 = \rho(A)$ and $\mathcal{K}_2(X) = 1$), this estimation is not really useful, first because the condition number $\mathcal{K}_2(X)$ of the matrix X is generally unknown, then because it may be very large.

2.1.8.4 BiConjugate Gradient algorithm

While the Conjugate Gradient method is designed for symmetric positive definite systems, the BiConjugate Gradient method can be applied to non-symmetric systems. This method is based on the Lanczos biorthogonalization; it consists in replacing the orthogonal sequence of residuals by two mutually orthogonal sequences. In counterpart, the minimization is not ensured anymore. The second conjugate gradient is provided by using A^T instead of using A. So there are two residual sequences which are defined by:

$$r^{(k+1)} = r^{(k)} - \alpha^{(k+1)} A p^{(k+1)}, \quad \tilde{r}^{(k+1)} = \tilde{r}^{(k)} - \alpha^{(k+1)} A^T \tilde{p}^{(k+1)} \tag{2.25}$$

where the search directions are defined as follows:

$$p^{(k+1)} = r^{(k)} + \beta^{(k)} p^{(k)}, \quad \tilde{p}^{(k+1)} = \tilde{r}^{(k)} + \beta^{(k)} \tilde{p}^{(k)} \tag{2.26}$$

In order to have the biorthogonality relations

$$(\tilde{r}^{(i)}, r^{(j)}) = (\tilde{p}^{(i)}, A p^{(j)}) = 0 \quad \text{if } i \neq j, \tag{2.27}$$

$\alpha^{(k+1)}$ and $\beta^{(k+1)}$ are defined as follows:

$$\alpha^{(k+1)} = \frac{(\tilde{r}^{(k)}, r^{(k)})}{(\tilde{p}^{(k+1)}, Ap^{(k+1)})}, \qquad \beta^{(k+1)} = \frac{(\tilde{r}^{(k+1)}, r^{(k+1)})}{(\tilde{r}^{(k)}, r^{(k)})} \qquad (2.28)$$

From the algorithmic point of view, the BiConjugate Gradient algorithm (Algorithm 2.12) presents some similarities with the Conjugate Gradient one. Nevertheless, it has the drawback of requiring the computation of the transpose product $A^T \times PTilde$. For some applications, the computation of this product is not possible. For example, it is not possible to assemble elements of the matrix for some applications.

Concerning the convergence of this algorithm few theoretical results are known. For symmetric positive definite systems, convergence results are similar to those of conjugate gradient, even if it requires twice the cost per iteration. For nonsymmetric matrices it has been proved that the method is more or less comparable to full GMRES (in terms of the number of iterations) [57].

2.1.9   Preconditioning

As we have seen in the previous sections, the speed of a minimization method to solve a linear system $Ax = b$ depends on the condition number $\mathcal{K}(A)$ of A. The objective of preconditioning is to solve an equivalent system whose condition number is as close as possible to 1.

This equivalent system has the form

$$M^{-1}Ax = M^{-1}b.$$

The principle of preconditioning techniques is to find an approximation of $M^{-1} = A^{-1}$ such that $\mathcal{K}(M^{-1}A)$ is the nearest possible to 1. Since $\mathcal{K}(A^{-1}A) = 1$, the best theoretical choice of M^{-1} is A^{-1}, but in practice, we have to choose an inexpensive approximation of A^{-1}.

The new linear system is then solved (when it is possible) by a minimization method giving rise to a preconditioned minimization algorithm. If we apply the gradient algorithm (Algorithm 2.19) with a fixed step p (Richardson algorithm) to a linear system $M^{-1}Ax - M^{-1}b = 0$, then we obtain,

$$\begin{cases} x^{(0)} \ given \\ x^{(k+1)} = x^{(k)} + p \left(M^{-1}b - M^{-1}Ax^{(k)} \right) \end{cases}$$

Then if we choose $M = D$, the diagonal part of A, then we have

$$\begin{cases} x^{(0)} \ given \\ x^{(k+1)} = x^{(k)} + p \left(D^{-1}b - D^{-1}Ax^{(k)} \right) \end{cases}$$

or in an equivalent way

$$\begin{cases} x^{(0)} \ given \\ Dx^{(k+1)} = Dx^{(k)} + p(b - Ax^{(k)}). \end{cases}$$

Algorithm 2.12 BiConjugate Gradient algorithm

Size: size of the matrix
A[Size][Size]: matrix
X[Size]: solution vector
R[Size]: residual vector
RTilde[Size]: second residual vector
P[Size]: search direction vector
PTilde[Size]: second search direction vector
Q[Size]: orthogonal vector to the search direction
QTilde[Size]: orthogonal vector to the second search direction
Alpha, Beta, Rho, RhoOld: scalar variables

$R \leftarrow B - A \times X$
Choose *RTilde*, for example *RTilde* = R
$i \leftarrow 1$
repeat
   $Rho \leftarrow (R, RTilde)$
   if $Rho = 0$ **then**
      method fails
   end if
   if i=1 **then**
      $P \leftarrow R$
      $PTilde \leftarrow RTilde$
   else
      $Beta \leftarrow Rho/RhoOld$
      $P \leftarrow R + Beta \times P$
      $PTilde \leftarrow RTilde + Beta \times PTilde$
   end if
   $Q \leftarrow A \times P$
   $QTilde \leftarrow A^T \times PTilde$
   $Alpha \leftarrow Rho/(PTilde, Q)$
   $X \leftarrow X + Alpha \times P$
   $R \leftarrow R - Alpha \times Q$
   $RTilde \leftarrow RTilde - Alpha \times QTilde$
   $RhoOld \leftarrow Rho$
   $i \leftarrow i + 1$
until stopping criteria is reached

Then we obtain the relaxed Jacobi algorithm to solve $Ax = b$. So, the Richardson algorithm preconditioned with the diagonal matrix of A coincides with the relaxed Jacobi algorithm.

In the following we will briefly describe the most used preconditioning techniques, namely Jacobi, SOR, SSOR and ILU preconditioning.

2.1.9.1   Jacobi, SOR, SSOR and ILU preconditioning

Consider a matrix A and its splitting

$$A = D - L - U$$

where D, L and U are, respectively, the diagonal, the lower and the upper parts of A.

The Jacobi preconditioning is given by the preconditioning matrix,

$$M_J = D.$$

The Gauss-Seidel preconditioning is defined by the preconditioning matrix

$$(D - L),$$

while the preconditioning matrix of the SOR preconditioning is

$$M_{SOR} = \frac{1}{\omega}(D - \omega L).$$

Thus, the Jacobi preconditioning matrix is simply the diagonal part of A and the Gauss-Seidel preconditioning matrix is the lower triangular part of A.

The SSOR preconditioning matrix is defined by ([52], [11])

$$M_{SSOR} = \frac{1}{\omega(2 - \omega)}(D - \omega L)D^{-1}(D - \omega U), \qquad (2.29)$$

where $\omega \in \,]0, 2[$.

2.1.9.2   Preconditioning matrices for the conjugate gradient algorithm

Suppose now that A is a symmetric positive definite matrix and split A into

$$A = D - L - L^T$$

Following (2.29), the SSOR preconditioning matrix is

$$M_{SSOR} = \frac{1}{\omega(2 - \omega)}(D - \omega L)D^{-1}(D - \omega L^T),$$

where $\omega \in \,]0, 2[$.

Note that M_{SSOR} does not require any computation nor any storage. Moreover the matrix M_{SSOR} can be written in the form

$$M_{SSOR} = C_\omega C_\omega^T,$$

where

$$C_\omega = \frac{(D - \omega L)D^{-1/2}}{(\omega(2 - \omega))^{1/2}},$$

which facilitates the resolution of the linear system $M_{SSOR}z = r$ (r being the residual) required at each step. If we choose $\omega = 1$, then we obtain the symmetric Gauss-Seidel iterations

$$M_{SGS} = (D - L)D^{-1}(D - L^T).$$

To obtain a preconditioned conjugate gradient algorithm, the matrix of the new linear system has to be symmetric. Since the product of symmetric matrices is not necessarily symmetric, the conjugate gradient algorithm cannot be directly applied to a matrix $M^{-1}A$ even if the matrix M^{-1} is symmetric.

If M^{-1} is positive definite, then there exists a matrix $M^{-1/2}$ such that $\left(M^{-1/2}\right)^2 = M^{-1}$. The system $M^{-1}Ax = M^{-1}b$ can be written

$$M^{-1}Ax = \left(M^{-1/2}\right)^2 b$$
$$M^{1/2}M^{-1}Ax = M^{1/2}\left(M^{-1/2}\right)^2 b$$
$$M^{1/2}M^{-1}Ax = M^{-1/2}b$$
$$M^{1/2}M^{-1}AM^{-1/2}M^{1/2}x = M^{-1/2}b.$$

Putting $y = M^{1/2}x$, we obtain the system

$$M^{-1/2}AM^{-1/2}y = M^{-1/2}b.$$

The matrix $M^{-1/2}AM^{-1/2}$ being symmetric positive definite, we can apply the conjugate gradient algorithm to it.

2.1.9.3 Implementation of the preconditioned conjugate gradient solver

Using a preconditioner with a numerical algorithm often does not require many modifications in the algorithm. Nonetheless, the difficulty lies in the way of building the preconditioner. Iterative solvers are easily adaptable to be used with a preconditioner. For more information, interested readers are invited to read some of the numerous books dedicated to this topic. Roughly speaking the modifications for an iterative solver concern the use of a `Solve` procedure that can compute the inverse of the preconditioned matrix and multiply it by the residual vector. Later, the result of this `Solve` procedure is used in the algorithm. Algorithm 2.13 describes the preconditioned version of the Conjugate Gradient algorithm. Compared to the simple version, there are few differences.

Algorithm 2.13 Preconditioned Conjugate Gradient algorithm

Size: size of the matrix
A[Size][Size]: matrix
X[Size]: solution vector
R[Size]: residual vector
M[Size][Size]: preconditioned matrix
Z[Size]: solution vector of the system $MZ = R$
P[Size]: search direction vector
Q[Size]: orthogonal vector to the search direction
Alpha, Beta, Rho, RhoOld: scalar variables

R ← B−A×X
i ← 1
repeat
  Z ← Solve(M, R)
  Rho ← (R,Z)
  if i=1 **then**
    P ← Z
  else
    Beta ← Rho/RhoOld
    P ← Z+Beta×P
  end if
  Q ← A×P
  Alpha ← Rho/(P,Q)
  X ← X+Alpha×P
  R ← R−Alpha×Q
  RhoOld ← Rho
  i ← i+1
until stopping criteria is reached

Algorithm 2.14 gives the `Solve` procedure that uses the SSOR preconditioner. It is supposed to be used with the matrix A as a parameter for M in $Solve(M, R)$. This procedure is based on Equation (2.29). There are three loops in the algorithm which correspond to the three parts of the equation: $(D - \omega L)$, $\frac{D^{-1}}{\omega \times (2-\omega)}$ and $(D - \omega U)$. Of course the result of each part is used for the next one. The first and the last parts correspond to a triangular matrix solve.

For more details on SSOR, interested readers are invited to consult [8, 120, 28].

Algorithm 2.14 SSOR Solve(M,R)

Size: size of the matrix
Z[Size]: solution vector
Z2[Size]: intermediate vector
Tmp: intermediate variable
Omega: relaxation parameter

for i=0 to Size−1 **do**
  Tmp← 0
  for j=0 to i−1 **do**
    Tmp ← Tmp+M[i][j]×Z2[j]
  end for
  Z2[i]← (R[i]−Omega×Tmp)/M[i][i]
end for
for i=0 to Size−1 **do**
  Z2[i]← Z2[i]×M[i][i]×(Omega×(2−Omega))
end for
for i=Size−1 to 0 **do**
  Tmp← 0
  for j=Size−1 to i+1 **do**
    Tmp ← Tmp+M[i][j]×Z[j]
  end for
  Z[i] ← (Z2[i]−Omega×Tmp)/M[i][i]
end for
return Z

2.1.9.4   Incomplete LU factorization

This method commonly called *ILU* is based on the well-known direct method LU [29]. The LU method is known to be very efficient for solving a system as soon as the factorization has been achieved. Unfortunately, this process is often complex to implement and time consuming compared to an iterative method to solve a sparse system. Another drawback of the LU method with a sparse linear system is that the factorization process tends to fill the matrix elements. Several techniques have been developed to circumvent this problem. For more details on direct methods and incomplete factorization techniques, interested readers are invited to consult [45, 46, 85, 112, 9, 28, 10].

In opposition to the traditional factorization method, the incomplete factorization ignores some elements that would be filled with a complete factorization. After the incomplete factorization process, we obtain a matrix $M = LU$ where L is a lower triangular matrix and U is an upper triangular one. The more the matrix M^{-1} approximates A^{-1}, the more the preconditioner is efficient.

The simplest form of the ILU algorithm builds an incomplete LU factorization without filling any empty element of the matrix. So, the factorization method needs to know the non-null elements of the matrix. In Algorithm 2.15, we need to know if elements of the matrix A are null or not, that is why we test them before applying the classical LU factorization. Consequently, only non-null elements are modified for the factorization. Hence, the ILU factorization algorithm is based on the standard LU factorization and only differs from it by not filling the empty elements of the matrix.

Algorithm 2.15 ILU factorization

Size: size of the matrix
M[Size][Size] : matrix used for the factorization, it must be initialized with the elements of the matrix to factorize

for i=1 to Size−1 **do**
   for k=0 to i−1 **do**
      if M[i][k] $\neq$ 0 **then**
         M[i][k] $\leftarrow$ M[i][k]/M[k][k]
         for j=k+1 to Size−1 **do**
            if M[i][k] $\neq$ 0 **then**
               M[i][j] $\leftarrow$ M[i][j]−M[i][k]×M[k][j]
            end if
         end for
      end if
   end for
end for

The ILU `Solve` procedure used in any preconditioned algorithm is similar to the LU `Solve` one. It consists in solving two triangular systems. Algorithm 2.16 illustrates this.

More efficient, but also more complex, preconditioners based on LU decomposition and other techniques are for example detailed in [29, 102].

2.2   Nonlinear equation systems

Nonlinear systems arise in multiple domains of computer science. Scientists are faced with nonlinear systems if they want to solve multiple optimization problems, root finding problems and many more. Nonlinear problems are far more difficult to solve than the linear ones.

Algorithm 2.16 ILU Solve(M,R)

Size: size of the matrix
Z[Size]: solution vector
Tmp: intermediate variable

for i=0 to Size−1 **do**
  Z[i] ← R[i]
  for j=0 to i−1 **do**
    Z[i] ← Z[i]−M[i][j]×Z[j]
  end for
end for
for i=Size−1 to 0 **do**
  Tmp← 0
  for j=Size−1 to i+1 **do**
    Tmp ← Tmp+M[i][j]×Z[j]
  end for
  Z[i] ← (Z[i]−Tmp)/M[i][i]
end for
return Z

Many books are dedicated to nonlinear systems, see, e.g., [93], [79] and the references therein. In this section we do not pretend to be exhaustive and we present the most commonly used method to solve nonlinear systems: the Newton-Kantorovich method (or the Newton method, for simplicity); see [77], [78], [79], [93]. Let us first review the basic concepts of nonlinear mappings.

2.2.1 Derivatives

Recall that a real function f of a single variable is differentiable at x_0 if there exists a real $a = f'(x_0)$ such that

$$\lim_{h \to 0} \frac{f(x_0 + h) - f(x_0) - ah}{h} = 0.$$

In the case of the n-dimensional real space, this definition is extended as follows

DEFINITION 2.3 *A nonlinear operator $F : D(F) \subset \mathbb{R}^n \to \mathbb{R}^m$ is Fréchet-differentiable at an interior point x of D if there exists a linear operator $A :$ such that for any $h \in \mathbb{R}^n$,*

$$\lim_{\|h\| \to 0} \frac{\|F(x_0 + h) - F(x_0) - Ah\|}{\|h\|} = 0. \tag{2.30}$$

The linear operator A ($m \times n$ matrix) is called the derivative of F at x_0 and is denoted by $F'(x_0)$.

If $f_1, ..., f_m$ denote the components of F and $\partial f_i(x)/\partial x_j$, the partial derivative of f_i at x_j, then $F'(x)$ is given by the Jacobian matrix

$$F'(x) = \begin{pmatrix} \partial f_1(x)/\partial x_1 & \cdots & \partial f_1(x)/\partial x_n \\ \vdots & \ddots & \vdots \\ \partial f_m(x)/\partial x_1 & \cdots & \partial f_m(x)/\partial x_n \end{pmatrix}.$$

The partial derivatives are also denoted by

$$J_{i,j} = \frac{\partial f_i(x)}{\partial x_j},$$

and so the Jacobian $F'(x)$ is denoted by $F'(x) = J(x) = (J_{i,j})_{1 \le i \le m;\ 1 \le j \le n}$.
If $F : D(F) \subset \mathbb{R}^n \to \mathbb{R}^n$ then $F'(x)$ is represented by the row vector

$$F'(x) = (\partial F(x)/\partial x_1, \cdots, \partial F(x)/\partial x_n),$$

and the column vector $(F'(x))^T$ is called the *gradient* of F at x and is denoted ∇F,

$$\nabla F = \begin{pmatrix} \partial F(x)/\partial x_1 \\ \vdots \\ \partial F(x)/\partial x_n \end{pmatrix}.$$

DEFINITION 2.4   *A vector valued function $F : \mathbb{R}^n \to \mathbb{R}^m$ is called continuously differentiable if F' is continuous. This means that each component f_j of F is differentiable and f_j' is continuous.*

2.2.2   Newton method

Initially, the Newton method allows us to find the roots of f (x such that $f(x) = 0$), where f is a continuously differentiable function of $\mathbb{R}$ into itself. It is expressed by

$$x^{(k+1)} = x^{(k)} - \frac{f(x^{(k)})}{f'(x^{(k)})} \tag{2.31}$$

This method is also called the *tangent method*. Each iterate $x^{(k+1)}$ is actually obtained using the previous iterate and drawing the tangent to the function f at point $(x^{(k)}, f(x^{(k)}))$ and taking its intersection with the x-axis, as illustrated in Figure 2.3.

The previous method is generalizable to $\mathbb{R}^n$. Let F be a nonlinear function of $\mathbb{R}^n$ into itself and x a vector of size n. Assume that equation

$$F(x) = 0 \tag{2.32}$$

has at least one solution x^* and that F' exists. Then the Newton method is defined by

$$x^{(k+1)} = x^{(k)} - F'(x^{(k)})^{-1} F(x^{(k)}) \tag{2.33}$$

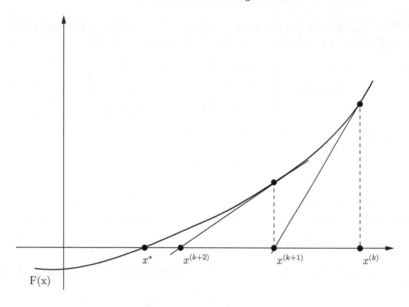

FIGURE 2.3: Illustration of the Newton method.

So it is possible to write the equivalent equation

$$F'(x^{(k)})x^{(k+1)} = F'(x^{(k)})x^{(k)} - F(x^{(k)}) \tag{2.34}$$

which can be expressed as

$$F'(x^{(k)})(x^{(k+1)} - x^{(k)}) = -F(x^{(k)}) \tag{2.35}$$

Let us denote by $\delta x^{(k+1)} = (x^{(k+1)} - x^{(k)})$ and by $J(x^{(k)})$ the Jacobian of F at $x^{(k)}$, then the previous equation can be rewritten as

$$J(x^{(k)})\delta x^{(k+1)} = -F(x^{(k)}) \tag{2.36}$$

The Newton method linearizes the nonlinear function and requires the resolution of a linear system at each iteration. A direct method or an iterative one may be used to solve the linear system obtained by the method. Because the computation of the Jacobian is time consuming, there exist several variants of the Newton method called *modified Newton methods*. For example, it is possible to only compute the linear system at some iterations or, in some particular cases, only at the first iteration. The algorithm is then called *quasi-Newton*. The number of iterations to reach the convergence may be increased but the benefit from not computing the Jacobian matrix is interesting.

Algorithm 2.17 illustrates the Newton algorithm or the quasi-Newton variant according to the fact that the Jacobian matrix is computed at each Newton iteration or not.

Algorithm 2.17 Newton and quasi-Newton algorithm

Size: size of the matrix
J[Size][Size]: Jacobian matrix
X[Size]: solution vector
Delta: residual vector

initialize X to the initial guess
repeat
   if first iteration or required **then**
      Computation of the Jacobian matrix J
   end if
   Delta $\leftarrow$ linearSolve($J, -F(X)$)
   X $\leftarrow$ X+Delta
until stopping criteria is reached

2.2.3   Convergence of the Newton method

The Newton algorithm (Algorithm 2.33) can be written as

$$x^{(k+1)} = T(x^{(k)}), \tag{2.37}$$

where

$$T(x^{(k)}) = x^{(k)} - F'(x^{(k)})^{-1} F(x^{(k)}).$$

In what follows, we shall give local and global convergence results on Newton methods.

THEOREM 2.9

Let x^* be a solution of Equation (2.32). Assume that $F'(x^*)$ has a bounded inverse and that $F'(x)$ is Lipschitz continuous with constant L in some neighborhood of x^*

$$\|F'(x) - F'(y)\| \le L \|x - y\|, \tag{2.38}$$

then the iteration vectors generated by the Newton algorithm satisfy

$$\left\| x^{(k+1)} - x^* \right\| \le c(\varepsilon, x^{(0)}) \varepsilon^{2n},$$

where $x^{(0)}$ is the initial vector iterate which is assumed to be sufficiently close to x^* and ε is any arbitrary small number. $c(\varepsilon, x^{(0)})$ is a number which depends on ε and the initial guess $x^{(0)}$.

PROOF   see [79].   □

In the following we give a global convergence condition on the Newton algorithm.

Let $x^{(0)}$ be the initial iterate of the Newton algorithm. Suppose that F is defined and differentiable on a ball $B(x^{(0)}, R)$ of radius R and center $x^{(0)}$. Suppose also that $F'(x)$ satisfies the Lipschitz condition (2.38). We assume that $\left(F'(x^{(0)})\right)^{-1}$ is well defined. The following theorem is proved in [79].

THEOREM 2.10
Assume that

$$\left\|\left(F'(x^{(0)})\right)^{-1}\right\| \le b_0 \text{ and } \left\|\left(F'(x^{(0)})\right)^{-1} F(x^{(0)})\right\| \le \eta_0 \text{ and } h_0 = b_0 L \eta_0 \le \frac{1}{2}.$$

If

$$R \ge r_0 = \frac{1 - \sqrt{1 - 2h_0}}{h_0} \eta_0$$

then the iterations generated by the Newton algorithm converge to a solution x^* of Equation (2.32) in the ball $B(x^{(0)}, r_0)$.

In the quasi-Newton algorithm $F'(x^{(k)})$ is replaced by the constant value $F'(x^{(0)})$, so this algorithm can be written as

$$x^{(k+1)} = T x^{(k)}.$$

The general convergence results of Theorems 1.2 and 1.3 of Section 1.2 can be applied to the quasi-Newton algorithm. We shall give below an analogue of the above theorem for the quasi-Newton algorithm.

Suppose that F is defined and Fréchet differentiable on a ball $B(x^{(0)}, R)$ in which the derivative $F'(x)$ satisfies the Lipschitz condition (2.38). Assume also that $\left(F'(x^{(0)})\right)^{-1}$ is well defined. Let $\left\|\left(F'(x^{(0)})\right)^{-1}\right\| \le b_0$ and $\left\|\left(F'(x^{(0)})\right)^{-1} F(x^{(0)})\right\| \le \eta_0$, then

THEOREM 2.11
If

$$h_0 = b_0 L \eta_0 < \frac{1}{2}$$

and

$$R \ge r_0 = \frac{1 - \sqrt{1 - 2h_0}}{h_0} \eta_0,$$

then the iterations generated by the quasi-Newton algorithm converge to a solution $x^* \in B(x^{(0)}, r_0)$ of Equation (2.32).

2.3   Exercises

1. Prove the equalities (1.1), (1.2) and (1.3) of Chapter 1.

2. *Norm equivalence theorem.* Let $\| \cdot \|$ and $\| \cdot \|^1$ be any two norms on $\mathbb{R}^n$. Prove that there exist constants $c_2 \geq c_1 > 0$ such that

$$c_1 \|x\| \leq \|x\|^1 \leq c_2 \|x\|, \ \forall x \in \mathbb{R}^n.$$

3. *Neumann Lemma.* Prove that if A is a $n \times n$ matrix such as $\rho(A) < 1$, then $(I - A)^{-1}$ exists and

$$(I - A)^{-1} = \lim_{k \to +\infty} \sum_{i=0}^{k} A^i.$$

4. Prove that if $\|A\| < 1$, then $I - A$ is nonsingular and

$$\left\| (I - A)^{-1} \right\| \leq \frac{1}{(1 - \|A\|)}.$$

5. Let $A = (A_{i,j})_{1 \leq i,j \leq n}$ be a square matrix; prove that $N(A) = \sum_{i,j=1}^{n} |A_{i,j}|$ defines a matrix norm.

6. Prove that $M(A) = \max_{1 \leq i,j \leq n} |A_{i,j}|$ does not define a matrix norm.

7. Prove that $N(A) = n(\max_{1 \leq i,j \leq n} |A_{i,j}|)$ defines a matrix norm.

8. Let A be a positive definite matrix. Prove that $N(x) = \sqrt{x^T A x}$ is a vectorial norm.

9. Prove the equivalence

$$\lambda \text{ is an eigenvalue of } A \Longleftrightarrow 1/\lambda \text{ is an eigenvalue of } A^{-1}$$

10. Let $A = (A_{i,j})_{1 \leq i,j \leq n}$ be a square matrix such that $\forall i, j \in \{1, ..., n\}$, $|A_{i,j}| < 1$. Prove that $\rho(A) < 1$.

11. ([113]) Consider the matrix $A = \begin{pmatrix} 5 & 2 & 2 \\ 2 & 5 & 3 \\ 2 & 3 & 5 \end{pmatrix} = D + L + U$ where D, L and U are, respectively, the diagonal, the lower triangular and the upper triangular matrices of A.

    (a) Prove that A is positive definite and compute the spectral radius ρ of the by-point Gauss-Seidel matrix $-(D + L)^{-1}U$.

(b) Compute the spectral radius of the per-block Gauss-Seidel matrix
$-(\mathcal{D}+\mathcal{L})^{-1}\mathcal{U}$ where $\mathcal{D} = \begin{pmatrix} 5 & 2 & | & 0 \\ 2 & 5 & | & 0 \\ 0 & 0 & | & 5 \end{pmatrix}, \mathcal{L} = \begin{pmatrix} 0 & 0 & | & 0 \\ 0 & 0 & | & 0 \\ 2 & 3 & | & 0 \end{pmatrix}$ and $\mathcal{U} = \mathcal{L}^T$

and compare it to ρ.

12. Prove that an irreducible diagonally dominant matrix is nonsingular.

13. *Gerschgorin Circle theorem.* Let $A = (A_{i,j})_{1 \le i,j \le n}$ be a real or complex square matrix and define the set

$$G = \bigcup_{i=1}^{n} \left\{ z, \ |a_{i,i} - z| \le \sum_{j \ne i} |a_{i,j}| \right\},$$

Prove that every eigenvalue of A lies in G.

14. Consider a contractive fixed point mapping T whose domain of definition is a complete metric space E. Let x^* denote the fixed point of T. Let us fix some $u^{(0)} \in E$. Denote by l the constant of contraction of T. Define the following sets

$$E^{(k)} = \left\{ u \in E, d(u, x^*) \le l^k d(u^0, x^*) \right\}.$$

Show that the sequential iterative algorithm defined by

$x^{(0)} \in E^{(0)}$
for k = 0,1,... **do**
$\quad x^{(k+1)} \leftarrow T(x^{(k)})$
end for

converges to x^*.

15. Consider the problem (1.5) with the condition (1.6) of Chapter 1.

   (a) Show that the matrix A of the linear system (1.7) is positive definite and that $A^{-1} > 0$.

   (b) Write the Jacobi algorithm to solve this problem.

   (c) Write the Gauss-Seidel algorithm.

   (d) Write the Successive OverRelaxation (SOR) algorithm and compute the optimal relaxation parameter.

   (e) By taking $a = b = 0$, $f(x) = 2\sin(x) + 1$ and $n = 15$, plot on the same graphic the solutions obtained by the three algorithms.

16. Consider again the problem (1.5) with the conditions (1.6) of Section 1.3.

(a) Find a block decomposition of A of the form

$$A = D - L - U,$$

where D is a positive definite block diagonal matrix and $-L$ and $-U$ are the block lower and upper parts of A and such that the $2D - A$ is a positive definite matrix.

(b) Show that the block Jacobi and the SOR methods associated with the above decomposition converge.

17. Show that the mapping $T : [0, 1] \subset \mathbb{R} \to \mathbb{R}$ defined by $T(x) = \frac{1}{2}x + 2$ is contractive but has no fixed point.

18. Write a program that solves the equation

$$\frac{e^x}{2} - \cos(x) = 0,$$

with the initial guess $x^{(0)} = 20$.

19. Let F be a nonlinear mapping from $\mathbb{R}^n$ into itself which has the decomposition

$$F(x) = Ax - H(x), \tag{2.39}$$

where A is a nonsingular $n \times n$ matrix and H is a nonlinear mapping.

(a) Write a program that allows the computation of the iterates

$$x^{(k+1)} = x^{(k)} - A^{-1}F(x^{(k)}) \text{ and } x^{(0)} \text{ given.} \tag{2.40}$$

(b) What is the condition of convergence of the iterations (2.40)?

(c) Give another simple formulation of the iterations (2.40).

(d) Find an example of nonlinear mapping F satisfying (2.39) such that (2.40) locally converges.

20. Consider the Newton algorithm

$$x^{(k+1)} = x^{(k)} - F'(x^{(k)})^{-1}F(x^{(k)}). \tag{2.41}$$

The Newton-SOR method consists in solving the linear problem involved in (2.41) by the SOR algorithm at each Newton iteration.

(a) Write the Newton-SOR algorithm by partitioning $F'(x^{(k)})$ into $D^{(k)} - L^{(k)} - U^{(k)}$.

(b) Write the two-stage algorithm corresponding to the Newton-SOR algorithm.

21. Several methods have been defined to find roots of polynomial. The most popular methods are the Newton method, the Aberth method [34] and the Durand-Kerner method [39]. A polynomial P of size n with complex coefficients has the following form:

$$P(z) = \sum_{i=0}^{n} a_i z^{n-i} \tag{2.42}$$

with $a_0 = 1$, $a_n \neq 0$ and $a_i \in \mathbb{C}$.

The Durand-Kerner method, which allows us to find every root z_i of the polynomial P, is defined by

$$z_i^{(k+1)} = \frac{P(z_i^{(k)})}{\displaystyle\prod_{j=1, j\neq i}^{n} (z_i^{(k)} - z_j^{(k)})} \tag{2.43}$$

for all $i \in [1, n]$.

The Aberth method, which allows us to find every root z_i of the polynomial P, is defined by

$$z_i^{(k+1)} = z_i^{(k)} - \frac{\dfrac{P(z_i^{(k)})}{P'(z_i^{(k)})}}{1 - \dfrac{P(z_i^{(k)})}{P'(z_i^{(k)})} \displaystyle\prod_{j=1, j\neq i}^{n} \dfrac{1}{(z_i^{(k)} - z_j^{(k)})}} \tag{2.44}$$

for all $i \in [1, n]$.

Design an iterative algorithm that computes the root of a polynomial using those methods.

Chapter 3

Parallel Architectures and Iterative Algorithms

Introduction

As seen in the previous chapter, iterative methods can be used on a large class of numerical problems. However, in numerous scientific applications, the size of the problem and/or the amount of required computations implies the use of a parallel system. Unfortunately, or fortunately, there is neither a single kind of parallel system nor a single kind of parallel iterative algorithm. Hence, the subject of this chapter is twofold: the first goal is to present the most common kinds of parallel architectures which can be encountered throughout the world and the second goal is to provide a classification of the parallel iterative algorithms.

A brief review of the evolution of parallel systems is given in Section 3.1. Then, Section 3.2 presents the classical parallel architectures and the main features which differentiate them. In Section 3.3, the trends of used configurations are discussed. Finally, a classification of parallel iterative algorithms is proposed in Section 3.4 with a focus on their respective advantages and drawbacks.

3.1 Historical context

The concept of parallelism is not new and was already extensively used far before the emergence of computer science. That concept is quite simple since it consists in gathering several working units and making them collaborate to perform a given task. Obviously, that definition is very broad and thus holds for many systems which are not in the scope of this book. However, it is exactly that concept which is used in computer science and it encompasses all the computing parallel systems going from parallel machines to distributed clusters while also including pipelines.

Since the beginning of parallel computing in the mid fifties, the evolution of parallel systems has been influenced by several factors: the progress of the interconnection networks and of the Integrated Circuits (IC) technology, but also the decrease in the production costs of processing units. In fact, the improvements of the networks have tended to increase the distance between the processing units whereas the advances related to ICs and to the financial aspect have influenced their nature.

In this way, parallelism initially invaded computers at the processor level under several aspects. The first one took place during the era of scalar processors, in the development of coprocessors taking in charge some specific tasks of the working unit (mathematical operations, communications, graphics,...) and relieving the Central Processing Unit (CPU). Another aspect has resided in the processor itself. The development of the Complementary Metal-Oxide-Semiconductor (CMOS) technology since 1963 and of the Very-Large-Scale Integration (VLSI) since the 1980s have allowed the inclusion of more and more complex components in the processors such as pipelines and multiple computation units. More recently, as the CMOS technology is nearer and nearer its physical limits, that intrinsic parallelization of the processors has logically been followed by the emergence of multiple-core processors.

Except for that last development, those forms of parallelism are hidden to the programmer of the machine; this is why we call it intrinsic parallelism. Concerning the explicit one, requiring a specific programming, the first parallel systems were tightly coupled such as vector processors, typically composed of a collection of basic processors on the same IC board and either used in a Simple Instruction Multiple Data (SIMD) way or in a pipelined one. But, over time, the links between the processors have become longer and longer, going from the inner scale of the computers (central buses, crossbar switches...) to their outer scale (local networks) to finally end up at the highest scale of the Internet since the beginning of the 21st century. As previously mentioned, in conjunction with that phenomenon, processors have become more and more complex due to the progress in electronics. Nevertheless, it is more the financial aspect than the technological one which has oriented the design of the following parallel systems. As the cost of personal computers sharply decreased in the nineties while the cost of parallel machines continued to be very high (due to specific hardware design and build), the economical constraints led more and more scientific organizations to develop local clusters. Most of those clusters were initially homogeneous in the way that they were composed of the same kind of machines linked together by the same kind of network. But here again, the fast evolution of the machines together with the sharp decrease of their costs considerably increased the turnover of the machines in those clusters, often leading to heterogeneous local clusters. The emergence of heterogeneity in local clusters has also been greatly facilitated by the effort in the standardization of the communication protocols. Eventually, the improving performances of the networks in the Internet have led to the interconnection of local clusters scattered over distinct geographical sites.

Finally, it can be seen that the advances in the communication networks and in the economical evolution of computer hardware have led to a large set of existing parallel architectures. The most common ones are detailed in the following part with a focus on the main features characterizing each of them.

3.2 Parallel architectures

As there are many ways to build a parallel system, the number of possible configurations is also very large. However, some architectures provide better performance than others. Of course, the notion of performance of a parallel system is quite theoretical since it also depends on the kind of application which is used on it. In fact, it is quite obvious that some systems are better suited to some kind of applications. In this way, even if in most of the cases, the parallel systems are quite general and allow the use of any kind of application; the design of such a system is often directly related to its intended use. So, although numerous variants may be deduced from the main classes of parallel systems, giving an exhaustive list of them is not in the scope of this book and the reader should, for example, refer to [47, 107, 96] for further details. Our concern here is to list the main kinds of parallel systems and to point out which kind of PIAs is most suited on each of them.

3.2.1 Classifications of the architectures

As in most domains, there exist a lot of possible classifications of the parallel architectures depending on the used criteria. However, Flynn's taxonomy [54] is the one commonly accepted as a reference in the domain. In that classification, there are four classes of parallel systems:

Single Instruction Single Data (SISD): corresponds to scalar monoprocessor systems performing only one instruction at a time on a single data.

Single Instruction Multiple Data (SIMD): is the class of vector processors and more generally of systems with a large number of small computing units allowing the application of an instruction on several data at the same time.

Multiple Instruction Single Data (MISD): is the only class that has not yet led to real implementations since it supposes the simultaneous application of different instructions on the same data. It seems that the range of applications corresponding to that particular architecture is quite reduced.

Multiple Instruction Multiple Data (MIMD): are the systems capable
of performing multiple instructions on different data at the same time.

As can be seen, that classification is based upon the relationship between the
instructions and the manipulated data. In a sense, it is quite general and
does not reflect all the aspects of parallel systems. Hence, other criteria can
be taken into account such as the way the memory is used or the physical
radius of the system, i.e., the physical distance between the processing units
(PUs).

Concerning the memory, it can be either shared by the processors, all the
processors accessing the same memory, or distributed over them, each processor owning its private memory with an exclusive access. Obviously, that
distinction is not relevant for the SISD systems and is not of great interest for
the MISD class which is not actually used. However, it is of concern to both
the SIMD and MIMD classes.

According to the radius of the system, another classification of the computer
architectures can be deduced which better reflects the evolution trends:

Monoprocessor Machines (MM): mainly representing the SISD computers available in the mass-market (PCs, workstations). They also include
SIMD machines containing one vector processor.

Parallel Machines (PM): built as a single machine containing several processing units. They include SIMD and MIMD architectures and potential combinations of them.

Local Clusters (LC): collections of independent computers gathered in the
same place and connected via a local network. Although they are intrinsically MIMD oriented, SIMD machines can be used at the node
level.

Distributed Clusters (DC): collections of local clusters scattered all
around the world and linked together via the Internet. Here also, those
systems mainly follow the MIMD model but they can include SIMD
parts.

The links between those two classifications are depicted in Figure 3.1.

It can be noticed that the MM class tends to be progressively replaced by
small sized PMs (2 or 4 cores). Moreover, since that first class of machines is
not directly in the scope of that part, only the last three classes are detailed
in the following.

3.2.1.1   Parallel machines

The concept of parallel machines is to include several PUs in the same machine running one instance of an operating system (OS). Hence, the common
architecture of most of the parallel machines can be seen as a collection of processing units linked together by a very fast interconnection network, partially

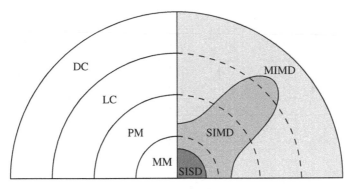

FIGURE 3.1: Correspondence between radius-based and Flynn's classification of parallel systems.

or completely implemented on ICs to provide bandwidths of the same order as the internal bandwidths of the nodes and very small latencies. Technically, that network may be implemented in different topological and physical ways from one machine to another (central buses, crossbars, butterflies or hypercubes...).

When the first researches on parallel systems began in the mid fifties, the advances in communication networks and in processor design only permitted the consideration of the gathering of a few simple processing units (PUs) in a single machine. The advances in both those domains have then led to an increase in the number of units in the machines and then to a more and more distinct separation of those units.

We present here two representative models of parallel machines which have been used for many years and which point out that progression toward larger radius systems. The distinction is made according to the memory management.

3.2.1.1.1 Parallel machines with shared memory The parallel machines with shared memory are a good example of the integrated approach since the processing units share some indispensable resources (memory, I/O operations). Schematically, the memory of the machine is accessible by every PU via the interconnection network. A representation of such a machine is given in Figure 3.2.

That kind of machine knew great success in the second half of the 20th century. The most famous examples are probably the Cray series such as Cray-1 and 2, and Cray X-MP and Y-MP. Nevertheless, most of the major computer vendors have also developed their own series of that kind of machine, as well as small companies specializing in supercomputers, such as Convex or Alliant, but also some academic institutions like the University of Illinois (Illiac IV).

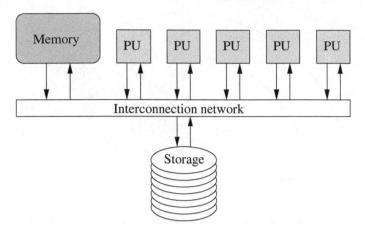

FIGURE 3.2: General architecture of a parallel machine with shared memory.

Although the production of that kind of machine has largely decreased in the last two decades, new machines are still being developed by some vendors as, for example, the Cray X1E.

The great advantage of such systems is that they neither require data distributions over the processors nor data messages between them. Moreover, the communications between PUs required for the control of the application are implicitly performed via the shared memory and can thus be very fast.

However, the centralization of the memory also presents some drawbacks. The first one is that it implies a very high memory bandwidth, potentially with concurrency, in order to avoid bottlenecks. Thus, the interconnection network between the memory and the PUs as well as the memory controller speed are often the limiting factors of the number of PUs included in that kind of machine. Moreover, there is also the problem of the concurrent accesses which may lead to incoherent results of the running application if not carefully managed. It is then necessary to use specific rules like mutual exclusion in order to ensure coherency. Those mechanisms often reduce the performance of the system.

3.2.1.1.2 Parallel machines with distributed memory This second version of parallel machines is very interesting since it is significant of the trend to make the PUs more and more independent. In this architecture, the PUs are still linked together by an interconnection network but the memory is no longer shared by all the PUs. Instead, each PU has its own memory with an exclusive access. However, in most cases, there remain some resources which are still shared like, for example, the mass storage or the I/O operations. Figure 3.3 depicts a typical distributed memory architecture.

That kind of machine knew great success and progressively replaced the

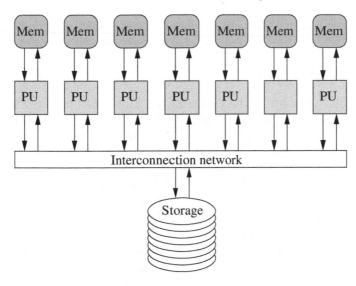

FIGURE 3.3: General architecture of a parallel machine with distributed memory.

previous type in the 1980s an 1990s. Logically, the major companies which built shared memory supercomputers have shown a great interest in that new type of architecture. Representative examples of such machines are the Cray T3D/T3E or the Intel iPSC/2 and Paragon. As before, small companies have also developed interesting products such as Connection Machines (CM-1 to CM-5) or MasPar (MP1 and MP2). Those systems have demonstrated their efficiency over the years and their larger flexibility in terms of design and configuration has made them more financially viable than their predecessors. This is probably why there are still important developments around them and they regularly appear among the most powerful parallel systems such as, for example, the IBM Blue Gene or the Cray Red Storm.

The advantages of such an architecture are directly related to the drawbacks of the previous one. In particular, the concurrent memory accesses are no longer an issue since only one PU accesses one memory bank. For the same reason, the problem of the memory bandwidth, although still important, is less critical. Hence, those systems are intrinsically better suited to include a very large number of PUs.

However, they also present some drawbacks. The most obvious ones are the necessity of a data distribution over the processors' own memory and the use of messages passing between processors to exchange data or information to control the application. The performances of the interconnection network between the PUs is also a critical point to ensure good performance. However, thanks to the larger flexibility in the network design induced by the distribution of the memory, different network topologies such as grids (2D or 3D), torus or crossbars have been developed to overcome that constraint.

3.2.1.2   Local clusters

The major advances in the communication networks which began in the eighties together with the sharp decrease in hardware costs have accelerated the radius growth of parallel systems. In some way, the clustering concept uses the opposite approach of the one used in parallel machines. Rather than designing an entire machine from scratch, the idea is to gather a collection of existing and independent PUs and to link them together via a local network. The obtained system can then be viewed as a single meta-machine and used in the same collaborative way as a parallel machine.

The implementations are of two sorts. The former, generally self-made, is the Beowulf approach introduced in the mid 1990s. The idea is to build low-cost parallel systems using only commercial off-the-shelf hardware and software. Such systems are also called *Networks Of Workstations* (NOWs) or *Clusters Of Workstations* (COWs). The latter, used by most vendors, consists in providing specific integration facilities (racks and cabinets) with optimized network and software environment. Whatever solution is used, the typical architecture of such clusters can be sketched as in Figure 3.4.

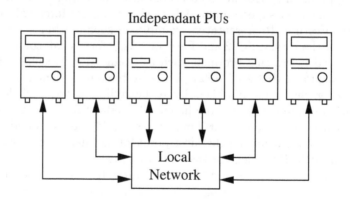

FIGURE 3.4: General architecture of a local cluster.

The major difference with the concept of parallel machines is that the PUs are not rigidly integrated in the machine. Hence, PUs can be easily added or suppressed from the system. Moreover, a distinction is usually made between clusters and constellations according to the relative number of nodes and processing units per nodes in the system. When the number of nodes is greater than the number of processing units per node, the system is considered as a cluster. The other case corresponds to constellations. That distinction is motivated by the impact on the programming methods used on those systems. Typically, message passing is intensively used on clusters which is not often the case on constellations.

The other big difference lies in the interconnection network which is often far slower than the fully integrated ones in parallel machines. Moreover, an additional weakness lies in the connections of the nodes themselves. Indeed, in most clusters, the connection of each node to the network is done through the connection bus of that node which often has restricted bandwidths and/or latencies. However, that difference tends to disappear with the latest generations of local communication networks and connection buses.

Numerous clusters of different sizes have been built during the last decade. The major reason for their success is that they represent an interesting alternative to the closed parallel machines previously described as they are generally cheaper while providing comparable theoretical peak performance. This is confirmed by the continuous increase of the proportion of clusters in the top 500 supercomputers (see [107]) since the mid 1990s. That proportion almost reaches three quarters of the June 2006 list. Nevertheless, it must be noted that most of those clusters are commercial products with integrated parts whereas self-made COWs only represent a tiny part.

Another important advantage is the flexibility of clusters compared to the rigidity of parallel machines which are often closed systems. The possibility of modifying the physical configuration of the system (PUs, network, RAM...) implies a larger convenience for its maintenance and a larger reactivity to technological advances. Indeed, defective or obsolete parts of a cluster can be easily replaced by new and better hardware. In this way, the computational power of a cluster can theoretically be in permanent evolution, following the progress in chip technology. Moreover, the flexibility even goes beyond the maintenance and upgrade aspects and offers scalability to those systems. The number of machines can be increased or decreased as needed. This enables us to potentially gather much more computational power and memory than in closed systems and thus to treat far more complex and/or large problems.

Of course, clusters are not perfect and their major weakness is obviously the network which remains slower than those integrated in parallel machines. However, as said above, that difference tends to disappear with the use of faster and faster local networks.

3.2.1.3 Distributed clusters/grids

The concept of distributed clusters (also referred to as grids) is quite a logical result of the great improvements in distant networks during the last few years combined with the stronger and stronger demand for more and more powerful computational systems. With the availability of quite fast distant networks, the use of several local clusters scattered on different geographical sites to solve a given problem becomes relevant. The principle of those distributed clusters is quite the same as that of local clusters, but just pushed a step further in physical radius. It consists of the interconnection of several local clusters via the Internet, as shown in Figure 3.5. That definition may seem a bit restrictive and a more general definition could explicitly include any kind of computers on the scattered sites (MM, PM or LC). In fact, the monoprocessor machines are implicitly included in the first definition since, although interest seems reduced, a local cluster may be reduced to a single MM. The problem is different with parallel machines. Usually, communications between the PUs of those machines and some external PUs are not supported. Moreover, their frequently imposed non-interactive use is an additional obstacle to the global coordination of the distributed cluster. This is why parallel machines have not been used in distributed clusters until now.

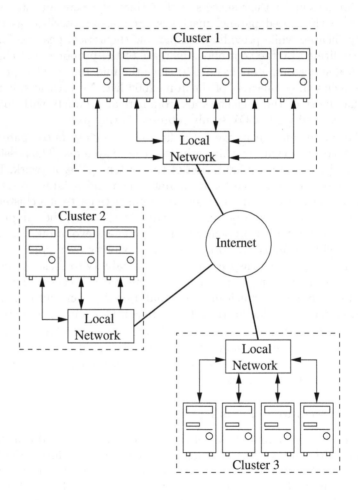

FIGURE 3.5: General architecture of a distributed cluster.

Examples of such systems are still very few, probably because of the important costs required to develop and maintain their global infrastructure (local clusters and high speed large scale networks) as well as their higher complexity of management compared to the rather specific application range they offer. However, national and/or international projects exist such as GRID'5000 in France and EGEE (Enabling Grid for E-sciencE) in Europe. Some software developments, like the SETI@home project at Berkeley University, have also led to the realization of a slightly different kind of distributed clusters. Their principle is based on cycle harvesting which consists in using some registered desktop computers linked to a central server via the Internet to perform a task given by the server when they are not locally in use. In such systems, there is no collaboration between the involved computers. Thus, their application range is even smaller than the collaborative distributed clusters discussed here.

The advantages of distributed clusters are quite obvious. They allow the gathering of a very large number of machines resulting in larger computational power and larger memory capacity. This scheme corresponds to the most powerful parallel systems since all the computers in the world could potentially be included in such a system via the Internet. Thus, the most complex and largest problems are likely to be treatable only on that kind of architecture.

However, at an even larger degree, those clusters suffer the same major drawback as their local counterparts. The communications between clusters are generally much slower than those inside local clusters which are themselves slower than those in parallel machines. So, the design of any application to be run on such a system must particularly focus on the logical organization of the processors according to the induced communication scheme, that is to say, the localization of the data exchanges in the network and their volume.

Moreover, the use of several distinct clusters implies another problem related to the security of computer systems. In most cases, the local networks connected to the Internet are protected from intrusions by filtering mechanisms such as firewalls. This implies that, contrary to the local clusters which have complete communication graphs, the distributed clusters often have non-complete ones. Indeed, local clusters usually have a single frontal access and the other machines inside it are not able to communicate with the exterior. Thus, a machine in a distributed cluster cannot directly communicate with all the other machines. This may be a strong constraint for some applications as a hierarchical communication mechanism must then be used. Such mechanisms are subject to bottlenecks and tend to slow down a bit more the communications in the parallel system.

3.3   Trends of used configurations

The previous classification depicts the general schemes of parallel systems which have been used since the beginning of parallelism. However, different combinations of those architectures have also arisen more recently. For example, the development of multi-core processors and bi- or quadri-processor boards have led to the mixing of shared and distributed memory. Indeed, a higher and higher number of parallel systems have a hierarchical architecture. That hierarchy is typically made of racks containing node cards containing mono/multi-core chips, as depicted in Figure 3.6. The best example of this probably is the IBM BlueGene/L (with one more level of hierarchy) whose one instance, installed at the Lawrence Livermore National Laboratory, has been at the top of the list of the most powerful supercomputers between 2005 and 2007 and probably will remain thus for still quite some time.

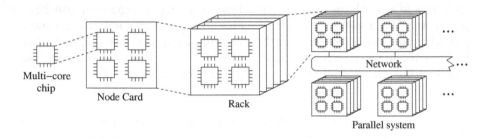

FIGURE 3.6: Hierarchical parallel systems, mixing shared and distributed memory.

Obviously, as more and more local clusters, including COWs, are composed of multi-core processors and/or multi-processor nodes, more and more distributed clusters logically belong to that class of mixed memory architectures. The level of hierarchy is just extended one level further. As already mentioned, the major differences with other hierarchical systems lie in the network properties at the outer level. However, it is worth noticing that the combination of shared and distributed memory is not systematic in distributed clusters. Also belonging to the class of hierarchical architectures are the S-COMA and ccNUMA systems. The main difference with the previous systems is that they work as shared memory systems.

Concerning the processors included in the latest systems, the trend goes toward a more and more intensive use of mass-market processors in supercomputers. Indeed, the strong competition between processor vendors has led to a drastic suppression of the series of processors which were not commercially viable. In the meantime, the increasing performance of mass-market processors and economical constraints have strongly induced the vendors to put mass-market processors in their supercomputers. Nevertheless, some vendors still continue to develop specific processors to put in supercomputers such as vector, massively multi-threaded or VLIW (Very Large Instruction Word) processors. Concerning those specific developments, it can be seen that after the invasion of the inner layers of the processors by parallelism, that one is currently resurging at the outer layers. There are some tries to re-use the parallel concepts usually taking place at the processor level at the scale of small groups of processors in order to design yet more powerful virtual processors. This is typically the case of IBM and its projects of Virtual Vector Architecture and cell processor.

Concerning the networks used, Gigabit Ethernet has been intensively used in non-integrated clusters and some integrated ones. However, it is rather penalized by its still high latencies. Other networks have emerged among them SCI, Infiniband or Myrinet. All those networks provide bandwidths of the order of the Gb/s and/or latencies the order of the microsecond. Moreover, they also often provide a large flexibility in the possible topologies.

Finally, according to current evolution, the major trend for future systems will certainly be the inclusion of heterogeneity at different levels of the parallel architectures. Some vendors, such as Cray or SGI, are already working on systems combining several kinds of processors (vector, scalar,...) and, in some sense, the networks are already heterogeneous in all the hierarchical architectures.

As a conclusion, the last remark that can be made is that the frontiers between the different parts of a parallel system are becoming less and less obvious. Processors tend to become mini multi-processor systems and clustering tends to be used at all the levels of multi-layer systems so that the term cluster alone becomes more and more inaccurate outside any specific context.

3.4   Classification of parallel iterative algorithms

The set of parallel iterative algorithms (PIA) being quite large, it contains algorithms which have a different global behavior and thus, different convergence conditions. So, it is necessary to distinguish some classes inside that set in order to precisely identify a given algorithm and to know what conditions hold on it.

As the global behavior of a PIA is mainly dictated by the synchronous/asynchronous nature of its iterations and communications, the following classification is based on those two criteria. Following that scheme, four classes should be expected. However, as the one with asynchronous iterations and synchronous communications is not relevant, there remain only three classes which are described here.

3.4.1 Synchronous iterations - synchronous communications (SISC)

This class corresponds to the most commonly used scheme in the PIA. At each iteration, each processor waits until it has received all the data computed at the previous iteration, coming from other processors before beginning its following iteration. This implies a synchronization of the iterations over the processors of the system. Moreover, the data exchanges are performed at the end of each iteration by synchronous global communications. An example of the execution flow of such an algorithm is given in Figure 3.7 in the case of two processors.

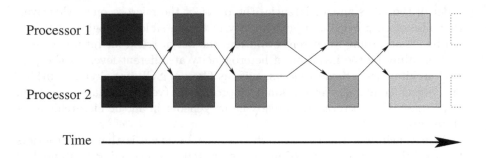

FIGURE 3.7: Execution flow of the SISC scheme with two processors.

Concerning the global behavior of this kind of algorithm, the synchronization of the iterations implies exactly the same sequence of iterations as in the sequential case. So, the global behavior of SISC algorithms is exactly the same as their sequential counterpart. This implies that their convergence conditions are identical to the ones standing for the SIA.

Those algorithms have known great success in numerical computation and are still widely used. The reasons for that success are manifold. The first one is that their performances are quite good on small physical radius parallel systems (parallel machines, local clusters) which are still the most commonly used. The second reason is that they do not require any further convergence analysis since their behavior is the same as in the sequential case. Finally, the last reason is that their design is quite straightforward starting from a sequential version and there exist several well-made synchronous communication libraries (PVM [66], MPI [71]) allowing non-specialists to write their own codes.

Unfortunately, the synchronous communications strongly penalize the performances of those algorithms, especially on systems with a slow and/or heterogeneous interconnection network. Indeed, as can be seen in Figure 3.7, there may be a lot of idle times (white spaces) between iterations (grey blocks) depending on the speed of the network used. So, although the synchronizations have a small impact in contexts of fast networks as in parallel machines and local homogeneous clusters, it is quite different in systems with slow or heterogeneous links between processors, which is typically the case in distributed clusters. In those last cases, the performance loss may be so important that the SISC algorithms cannot be reasonably used.

3.4.2 Synchronous iterations - asynchronous communications (SIAC)

This class has been developed in order to overcome the performance problem of the SISC algorithms. Its principle is to keep a synchronized iterative scheme while performing data exchanges between processors asynchronously. The synchronized iterations conserve the same global behavior as the SISC and thus the same convergence conditions as well. The asynchronous communications present the advantage of performing an overlapping of the computations with the communications.

In fact, before beginning a new iteration, each processor still waits for the data computed at the previous iteration, coming from other processors. However, the synchronous global communications are replaced by asynchronous sendings and blocking receipts of the data. In general, this scheme is associated with a flexibility of the communications. It consists in sending data or a group of data not at the end of the iteration but as soon as it has been updated during the iteration. This strategy relies on the assumption that the data sent during an iteration have a good probability of reaching their destination before the end of the current iteration. Hence, those data will be directly available on that destination for the next iteration and the processor will not have to wait for their delivery.

This yields a partial overlapping of the computations with the communications. It is partially due to the fact that neither the computation of the first data or group of data can be overlapped with communications nor the sending of the last group of data on the latest processor can overlap some computations on its destinations. Moreover, even the communications started during an iteration may be longer than this iteration, still implying a waiting time on the respective destinations. However, even if this scheme does not completely eliminate the idle times between iterations, the partial overlapping of computations with communications implies shorter idle times than those in the SISC and thus provides better overall performance. An example of the execution flow of the SIAC scheme is given in Figure 3.8 in the case of two processors. It can also be seen that the order of the communications may not be respected.

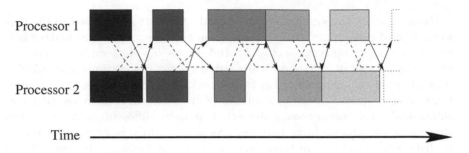

FIGURE 3.8: Execution flow of the SIAC scheme with two processors. In this example, the first half of data is sent as soon as updated (dashed arrows) and the second half is sent at the end of the iteration (solid arrows).

Concerning the global behavior of these algorithms, as each processor begins its next iteration as soon as it has received all its required data, all the processors may not begin their iterations at the same time. Nonetheless, for that same reason, a processor which would have finished its iteration before the others would not be able to begin the next iteration before them. Thus, it is not possible to have at any time during the whole iterative process two processors computing iterations with different numbers. In fact, at each time, the processors are either computing the same iteration (same number) or waiting for their required data. So, the notion of synchronism still holds in this scheme and the convergence conditions are the same as for the SISC and sequential algorithms.

3.4.3   Asynchronous iterations - asynchronous communications (AIAC)

Considering the previous classes, it appears that the last obstacle to a complete overlapping of the computations with the communications lies in the synchronization of the iterations. The principle used in the AIAC algorithms is to suppress that obstacle.

In that context, all the processors perform their iterations without taking care of the progression of the other processors. They do not wait for the latest updates of the data coming from the other processors but they keep on computing their iterations, using the version of those data they own at that time. That basic concept has been deduced from the models of Bertsekas [33] and El Tarazi [110]. As the processors do not wait for the data deliveries, there are no more idle times between the iterations, as can be seen in Figure 3.9 which depicts an example of the execution flow of the basic AIAC scheme for two processors.

Considering the global behavior of those algorithms, the asynchronism of

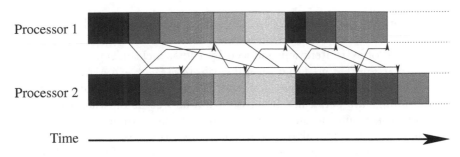

Time

FIGURE 3.9: Execution flow of the basic AIAC scheme with two processors. The horizontal part of the arrows represents the span of time between the data receipt and its integration in the computations.

the iterations implies that two processors may be computing different iterations at the same time (see Figure 3.9). In fact, the difference with the previous classes is even more complex than that and corresponds to the presence of indeterminism in the iterative process. Those mechanisms are detailed in Chapter 5 but intuitively, they correspond to the random evolutions during the process of the relative speeds of the processors and of the communication times between them.

Thus, although the set of fixed points or stabilization states stays the same between a synchronous PIA and its asynchronous version, their respective sequences of iterations may no longer be the same for any given initial state. So, the AIAC algorithms have a different global behavior than the algorithms of the previous classes implying different convergence conditions. Hence, they require a specific convergence analysis.

Another difference implied by the asynchronism is that the convergence requires at least the same number of iterations but often more than in synchronous algorithms. They are thus slower in terms of iterations. However, the penalties due to the synchronization of the iterations are generally so important that they largely compensate the time required to compute the extra iterations of the AIAC algorithm. So, depending on the parallel system used, it is possible to have a PIA which is faster than another while performing more iterations. The most favorable contexts are particularly the distributed clusters in which the communication links are often heterogeneous with some great differences between the links performances. The efficiency of the AIAC in those contexts comes from the fact that they are less sensitive to the communication delays and to their variations than the synchronous algorithms. Moreover, they also present some tolerance to the loss of data messages since such losses (in reasonable ratios) do not prevent the progression of the iterative process on both the sender and destination nodes. Both the processors keep on their computations and another message will be sent in a further iteration. However, that tolerance does not hold for the control messages used for the convergence detection and the halting procedure. Finally, in order to extend the efficiency of the AIAC to more contexts, it is essential to minimize the number of extra iterations induced by the asynchronism.

It is to attain that goal that several variants of the basic AIAC scheme presented in Figure 3.9 have been designed. Those variants depend on when the data sendings are performed in an iteration as well as on when the received data are taken into account in the computations. Among all the possible variants, two are particularly interesting; the former uses semi-flexible communications and the latter uses flexible communications. In opposition to the semi-flexible and flexible communications, the basic AIAC scheme previously presented is often referred to as the rigid communication model.

3.4.3.1 Semi-flexible communications

Concerning the semi-flexible schemes, there are in fact two symmetrical cases according to which side of the communication is flexible, the sender or the receiver.

3.4.3.1.1 Sender-side semi-flexibility In this first case, the receiver conserves the rigid scheme. So, all the data received during an iteration are integrated in the computations only at the beginning of the next iteration. On the opposite, the sender is flexible in the way that in place of sending all the required data at the end of each iteration, it uses the same sending policy as in the SIAC scheme. Hence, data are decomposed in groups which are asynchronously sent during the iteration as soon as they are updated. The goal of this method is to speed up the convergence of the process by making the most recent updates of the data available sooner on their destinations. The data groups may be reduced to a single element but generally they contain several ones in order to avoid the overloading of the network with a lot of small messages. Hence, particular attention must be paid to the design of those data groups in order to obtain the best progression speed-ups without decreasing the network performance too much.

The execution flow of this scheme is given in Figure 3.10.

3.4.3.1.2 Receiver-side semi-flexibility In this second case, it is the sender which conserves the rigid scheme and the receiver which becomes flexible. Hence, all the data required on another processor are asynchronously sent at the end of each iteration but as soon as they are received on their destination, they are integrated in the computations. This scheme is depicted in Figure 3.11.

The objective is also to speed up the convergence of the algorithm but it uses a different approach which consists in integrating the most recent data in the computations as soon as possible. Nonetheless, the direct integration of data during the computations is not always possible according to the kind

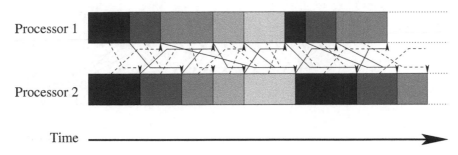

FIGURE 3.10: Execution flow of the sender-side semi-flexible AIAC scheme with two processors. In this example, the first half of data is sent as soon as updated (dashed arrows) and the second half is sent at the end of the iteration (solid arrows). The horizontal part of the arrows represents the span of time between the data receipt and its integration in the computations.

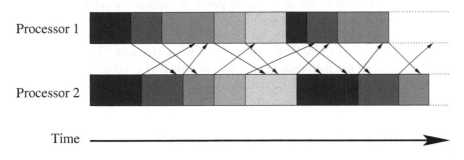

FIGURE 3.11: Execution flow of the receiver-side semi-flexible AIAC scheme with two processors.

of computations performed. So, the conditions of use of this scheme are more restrictive than those of the previous one.

3.4.3.2 Flexible communications

This scheme, introduced by Miellou, El Baz and Spitéri in [88] (see also [48, 63]), simply consists in a combination of the two semi-flexible schemes in order to take advantage of their respective improvements. So, on the sender side, the data are asynchronously sent by groups as soon as they are updated and, on the receiver side, the data are integrated in the computations as soon as they are received. The resulting execution flow is given in Figure 3.12.

This last scheme suffers the same restriction problem as the receiver-side semi-flexibility and the design of the data groups on the sender side must be carefully done. However, when it can be used, it generally yields better results than the two previous ones.

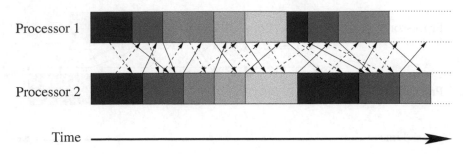

Processor 1

Processor 2

Time

FIGURE 3.12: Execution flow of the flexible AIAC scheme with two processors. In this example, the first half of data is sent as soon as updated (dashed arrows) and the second half is sent at the end of the iteration (solid arrows).

As said above, there are many more possible variants of AIAC algorithms. For example, it is possible to adapt the schemes presented here to a specific computing context or to a specific kind of application. The important point is that those adaptations will always deal with the way the asynchronous communications are managed on the sender side and on the receiver side.

In the following paragraph, the adequacy between the different kinds of PIAs and the parallel architectures previously described is detailed.

3.4.4   What PIA on what architecture?

3.4.4.1   Parallel machines

Concerning the use of PIAs on such systems, as they generally provide the fastest communications between processors, they are very well suited to synchronous algorithms.

More particularly, on shared memory machines, the sharing of the memory as well as the similarity of the PUs in those systems (in most of the cases) imply a drastic reduction of any potential asynchronism. So, although it is possible to implement PIAs without any explicit synchronization on such systems, their global behavior is likely to be almost the same as their synchronous counterparts.

Concerning the distributed memory machines, as in the previous case, it is still possible to design very fast interconnection networks with partial or complete integration on circuits. For that reason, this parallel context is also better suited to synchronous PIAs. But, as the communication delays are usually a bit larger than in shared memory systems, the use of asynchronous PIAs is a bit more relevant. Nonetheless, the extra iterations those algorithms often require compared to their synchronous versions are not likely to be compensated by the synchronization penalties which are rather reduced in that case.

3.4.4.2 Local clusters

In such computational contexts, the choice between synchronous and asynchronous PIAs starts becoming more ambiguous. In fact, as will be seen in Chapter 6, the synchronous versions are better suited for small homogeneous clusters with very fast networks whereas asynchronous algorithms provide better performance on large-size heterogeneous clusters. This comes from the fact that asynchronous algorithms are more robust to the heterogeneity of the processors and to slow communications. Indeed, when the number of processors increases, the network is likely to be much more solicited and the communications may be sharply slowed down by congestions. Also, when the processors have significant differences in performance, the speed of synchronous iterative processes is limited by the slowest processors which is not the case with their asynchronous counterparts. So, in those contexts, the asynchronous algorithms obtain better performance than the synchronous ones.

So, according to PIAs, that computational context is certainly the most versatile since both synchronous and asynchronous algorithms can be efficiently used.

3.4.4.3 Distributed clusters/grids

As mentioned above, asynchronous PIAs are more robust to slow communications and to heterogeneity than the synchronous ones. Hence, it is not a big surprise if they are the most suited PIAs to be run on distributed clusters. In fact, the communication constraint is generally so strong that synchronous algorithms (even outside the class of PIAs) must be prohibited on those kinds of systems. However, if future advances in communication networks provide much faster distant communications, synchronous algorithms may be reconsidered for use in such contexts. However, the problem of the heterogeneity of the processors remains and is even likely to be more important in those larger systems than in local clusters.

Chapter 4

Synchronous Iterations

Introduction

In this chapter, we are interested in parallel synchronous iterative algorithms for linear and nonlinear systems. Convergence results of the synchronous versions and their implementations are detailed.

We will concentrate on so-called *multisplitting algorithms* and their coupling with the Newton method. Multisplitting algorithms include the discrete analogues of Schwarz multi-subdomain methods and hence are very suitable for distributed computing on distant heterogeneous clusters. They are particularly well suited for physical and natural problems modeled by elliptic systems and discretized by finite difference methods with natural ordering.

The parallel versions of minimization like the methods exposed in Chapter 2 are not detailed in this chapter but it should be mentioned that, thanks to the multisplitting approach and under suitable assumptions on the splittings, these methods can be used as *inner iterations* of *two-stage* multisplitting algorithms.

4.1 Parallel linear iterative algorithms for linear systems

4.1.1 Block Jacobi and O'Leary and White multisplitting algorithms

Suppose that we have L processors $P_1, ..., P_L$ and that an unknown vector of dimension n is partitioned into L subvectors of dimensions n_i $(i \in \{1, ..., L\})$ so that $n = \sum_{i=1}^{L} n_i$, $\mathbb{R}^n = \prod_{i=1}^{L} \mathbb{R}^{n_i}$.

Consider the n-dimensional linear system

$$Ax = b, \; x \in \mathbb{R}^n, \tag{4.1}$$

and suppose that (4.1) has a unique solution x^*.

As seen in Section 2.1.6 of Chapter 2, block iterative algorithms can be deduced from by-point iterative algorithms by splitting the matrix A into $M - N$

where M and N are block matrices. The parallel block Jacobi algorithm consists in taking M as a block nonsingular diagonal matrix

$$
M = D = \begin{pmatrix} A_{11} & 0 & \cdots & \cdots & 0 \\ 0 & A_{22} & \ddots & & \vdots \\ \vdots & & \ddots & A_{33} & \ddots & \vdots \\ \vdots & & & \ddots & \ddots & 0 \\ 0 & \cdots & \cdots & 0 & A_{nn} \end{pmatrix} , \quad N = \begin{pmatrix} 0 & -A_{12} & -A_{13} & \ldots & -A_{1n} \\ -A_{21} & 0 & -A_{23} & \ldots & -A_{2n} \\ -A_{31} & -A_{32} & 0 & \ldots & -A_{3n} \\ \vdots & \vdots & \vdots & \ddots & \vdots \\ -A_{n1} & -A_{n2} & -A_{n3} & \ldots & 0 \end{pmatrix}
$$

$$(4.2)$$

where A_{ii} are matrices of dimension $n_i \times n_i$, so that at each iteration k, each processor P_i solves for $X_i^{(k+1)}$ the linear system

$$
A_{ii} X_i^{(k+1)} = - \sum_{j \neq i} A_{ij} X_j^{(k)} + B_i,
$$

where X_i and B_i are the ith block components of x and b of dimensions $n_i \times 1$ so that we have

$$
x = (X_1, ..., X_L)^T \text{ and } b = (B_1, ..., B_L)^T.
$$

The convergence of the parallel block Jacobi algorithm is deduced from Theorem 2.1 of Section 2.1 of Chapter 2. Indeed, it is sufficient to consider that the new fixed point mapping T is the one corresponding to the block Jacobi matrix $J = M^{-1}N$ where M and N are defined above in (4.2).

In this section we introduce O'Leary and White algorithms ([92], [118]) which generalize the parallel block Jacobi algorithms. Let us first recall some definitions and results which will be helpful in the comparison of the speed of convergence of the different forthcoming parallel algorithms.

DEFINITION 4.1   *We say that a vector x is nonnegative (positive), denoted $x \geq 0$ ($x > 0$), if all its entries are nonnegative (positive). A matrix B is said to be nonnegative, denoted $B \geq 0$, if all its entries are nonnegative. We compare two matrices $A \geq B$, when $A - B \geq 0$, and two vectors $x \geq y$ ($x > y$) when $x - y \geq 0$ ($x - y > 0$).*

DEFINITION 4.2   *Let A be a $n \times n$ real matrix. The decomposition $A = M - N$ is called a splitting if M is nonsingular. It is called a convergent splitting if $\rho(M^{-1}N) < 1$. A splitting $A = M - N$ is called:*
(a) regular if $M^{-1} \geq 0$ and $N \geq 0$
(b) weak regular if $M^{-1} \geq 0$ and $M^{-1}N \geq 0$.

THEOREM 4.1
Let $A = M - N$, where A and M are nonsingular square matrices.
Let $T = M^{-1}N$ and suppose that T is a nonnegative matrix, then

$$\rho(T) < 1 \Leftrightarrow A^{-1}N \geq 0.$$

Moreover

$$\rho(T) = \frac{\rho(A^{-1}N)}{1 + \rho(A^{-1}N)}.$$

PROOF   Suppose that $\rho(T) < 1$. Then

$$
\begin{aligned}
A^{-1}N &= \left[M(I - M^{-1}N) \right]^{-1} N \\
&= (I - T)^{-1}T \\
&= \sum_{p=1}^{\infty} T^p.
\end{aligned}
$$

As T is nonnegative, we deduce that $A^{-1}N$ is nonnegative.

Suppose now that $A^{-1}N \geq 0$. Then the Perron-Frobenius theorem implies that there exists a positive vector such that

$$Tx = \rho(T)x.$$

So

$$
\begin{aligned}
A^{-1}Nx &= (I - T)^{-1}Tx \\
&= \frac{\rho(T)}{1 - \rho(T)}x.
\end{aligned}
$$

As $A^{-1}N$ and x are nonnegative, the last equality implies that $\rho(T) < 1$.
Now, the equation just quoted above implies that

$$\frac{\rho(T)}{1 - \rho(T)} \leq \rho(A^{-1}N),$$

hence

$$\rho(T) \leq \frac{\rho(A^{-1}N)}{1 + \rho(A^{-1}N)}.$$

On the other hand, as $A^{-1}N \geq 0$, we have by the Perron-Frobenius theorem

$$
\begin{aligned}
Ty &= (I + A^{-1}N)^{-1}A^{-1}Ny \\
&= \frac{\rho(A^{-1}N)}{1 + \rho(A^{-1}N)}y,
\end{aligned}
$$

for some positive vector y, thus

$$\rho(T) \geq \frac{\rho(A^{-1}N)}{1 + \rho(A^{-1}N)}.$$

$\square$

THEOREM 4.2
Let $A = M - N$ be a weak regular splitting of A. Then the following assertions
are equivalent:

1. $A^{-1} \geq 0$.

2. $A^{-1}N \geq 0$.

3. $\rho(T) < 1$.

PROOF   1) implies 2) since $A^{-1}N = (I - T)^{-1}T = \sum_{p=1}^{\infty} T^p$ (Neumann
Lemma, see the Appendix).
2) $\Leftrightarrow$ 3) by Theorem 4.1.
3) $\Rightarrow$ 1) : since $A^{-1} = (I - T)^{-1}M^{-1} = \sum_{p=1}^{\infty} T^p M^{-1}$.   ☐

PROPOSITION 4.1
Consider a square $n \times n$ matrix A such that $A^{-1} \geq 0$. Let

$$A = M_1 - N_1 = M_2 - N_2$$

be two regular splittings of A. Denote by $T_1 = M_1^{-1}N_1$ and by $T_2 = M_2^{-1}N_2$,
then

$$N_2 \leq N_1 \Rightarrow \rho(T_2) \leq \rho(T_1),$$

so

$$R_\infty(T_1) \leq R_\infty(T_2).$$

PROOF   This is a consequence of Theorem 4.2 and the fact that the
function $f(x) = x/(1 + x)$ is monotone increasing.   ☐

The above results allow us to compare two block Jacobi like algorithms.
Indeed, the decompositions $A = M_1 - N_1$ and $A = M_2 - N_2$ give rise to the
block Jacobi algorithms whose iteration matrices are, respectively, $M_1^{-1}N_1$
and $M_2^{-1}N_2$. Indeed, the behaviors of synchronous iterations, generated by
these block Jacobi algorithms to solve the linear system (4.1), are, respectively,
described by the successive approximations associated to the fixed point mapping
$$T^{(1)} : \mathbb{R}^n \to \mathbb{R}^n$$
$$x \mapsto y = M_1^{-1}N_1 x + M_1^{-1}b$$

and
$$T^{(2)} : \mathbb{R}^n \to \mathbb{R}^n$$
$$x \mapsto y = M_2^{-1}N_2 x + M_2^{-1}b.$$

Then, Theorems 4.1 and 4.2 give sufficient conditions to ensure the convergence of block Jacobi algorithms. Theorem 4.1 allows us to compare the speed
of convergence of two given block Jacobi algorithms.

The multisplitting approach consists in partitioning the matrix A into horizontal band matrices. Then each processor, or group of processors, is responsible for the management of a band matrix and the associated unknown subvector of x. Multisplitting methods were first introduced by O'Leary and White in [92], [118], [116]; they define a multisplitting of A as a collection of L triplets (B_l, C_l, D_l) such that

1. $A = B_l - C_l$, for $l = 1, \ldots, L$ where B_l is nonsingular.

2. $\sum_l D_l = I$ where D_l $(l = 1, ..., L)$ are diagonal nonnegative matrices and I is the identity matrix.

   Then the multisplitting algorithm is defined as follows ($x^{(0)}$ given):

Algorithm 4.1 Multisplitting scheme

---
    for i=0,1,$\ldots$, until convergence **do**
        for l=1,$\ldots$,L **do**
            $y_l \leftarrow B_l^{-1}C_l x^{(i)} + B_l^{-1}b$
        end for
        $x^{(i+1)} \leftarrow \sum_l D_l y_l$
    end for

O'Leary and White [92] established the following result:

THEOREM 4.3
If for $l = 1, ..., L$, (B_l, C_l) are weak regular splittings of A satisfying $A^{-1} \geq 0$, then Algorithm 4.1 is convergent.

The convergence of O'Leary and White multisplitting algorithms given in Theorem 4.3 is based on Theorem 4.1. It can be seen that block Jacobi algorithms correspond to the particular case of the O'Leary and White multisplitting method where the matrix is partitioned into non-overlapping blocks and where the entries of the weighted diagonal matrices are null when they are not associated with the computation of the vector associated with the block diagonal matrix.

Since the work of O'Leary and White, several authors have studied multisplitting algorithms for linear and nonlinear systems; we refer to [59], [61], [60], [75], [27], [62], [5] and the references therein.

In the next section, we give a general formulation of multisplitting algorithms due to Bahi et al. [27]. This formulation allows us to put in the same theoretical framework, parallel block Jacobi algorithms, O'Leary and White multisplitting algorithms and the discrete analogues of parallel Schwarz algorithms.

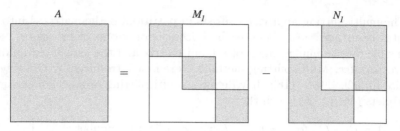

FIGURE 4.1: A splitting of matrix A.

4.1.2 General multisplitting algorithms

In this section we follow the general formulation of multisplitting algorithms given in [27]. These algorithms are described by the iterations generated by the successive approximations associated to an *extended fixed point mapping* defined from $(\mathbb{R}^n)^L$ into itself, where n is the dimension of the problem and L is the number of processors. This fixed point mapping is defined as follows:

$$\begin{cases} \mathcal{T} : (\mathbb{R}^n)^L & \longrightarrow \quad (\mathbb{R}^n)^L \\ X = (x^1, ..., x^L) & \longmapsto Y = (y^1, ..., y^L), \end{cases} \tag{4.3}$$

such that for $l \in \{1, ..., L\}$

$$\begin{cases} y^l = T^{(l)}(z^l) \\ z^l = \sum\limits_{k=1}^{L} E_{lk} x^k, \end{cases} \tag{4.4}$$

where E_{lk} are weighting matrices satisfying

$$\begin{cases} E_{lk} \text{ are diagonal matrices} \\ E_{lk} \geq 0 \\ \sum\limits_{k=1}^{L} E_{lk} = I_n \quad \text{(identity matrix)}, \quad \forall l \in \{1, ..., L\}. \end{cases} \tag{4.5}$$

In (4.4),

$$T^{(l)}(z^l) = M_l^{-1} N_l z^l + M_l^{-1} b \tag{4.6}$$

where

$$A = M_l - N_l, \quad l = 1, ..., L \tag{4.7}$$

is a splitting of A and M_l is, e.g., the block diagonal matrix defined in Figure 4.1.

Then it can be shown that if each splitting is convergent, i.e., if $\rho(M_l^{-1} N_l) < 1$, then the extended fixed point mapping is also convergent to the extended solution of (4.1), say $(x^*, ..., x^*)$, and then the synchronous algorithm converges. The convergence study will be detailed in the

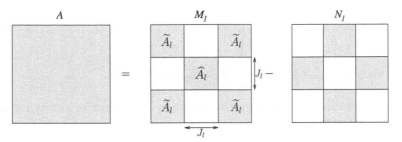

FIGURE 4.2: A splitting of matrix A using subset J_l of $l \in \{1, ..., L\}$.

next chapter in the more general context which includes the study of both synchronous and asynchronous algorithms.

In the following, a matrix A is partitioned as follows:

$$\left(\widehat{A_l}\right)_{i,j} = a_{i,j}, \ for \ i, j \in J_l,$$
$$\left(\widetilde{A_l}\right)_{i,j} = a_{i,j} \ for \ i, j \in J_l^C,$$
$$M_l = diag\left(\widehat{A_l}, \widetilde{A_l}\right),$$

where L denotes the number of processors and J_l are subsets of $\{1, ..., n\}$. The elements of J_l are indices of sub-components of a vector x of $\mathbb{R}^n$. To each $l \in \{1, ..., L\}$ we associate a splitting, so we obtain L splittings of A described in Figure 4.2.

We will now show how the extended fixed point defined above and the dependence of the weighting matrices on both l and k allow us to obtain particular standard algorithms such as O'Leary and White multisplitting algorithms.

The practical considerations on how to implement such algorithms are discussed in Section 4.4.5.

4.1.2.1 Obtaining O'Leary and White multisplitting

If the diagonal positive matrices E_{lk} depend only on k

$$E_{lk} = E_k$$

and satisfy

$$\begin{cases} \sum_{k=1}^{L} E_k = I_n \\ (E_k)_{i,i} = 0, \ \forall i \notin J_k \end{cases} \tag{4.8}$$

Then the synchronous iterations corresponding to O'Leary and White multisplitting are defined by the fixed point mapping (here $L = n$, $B_l = \mathbb{R}^n$),

$$T^{OW}(x^1, ..., x^L) = (y^1, ..., y^L) \ such \ that$$

$$\begin{cases} y^l = T^{(l)}(z) \\ z = \sum_{k=1}^{L} E_k x^k \end{cases}$$

where for $l \in \{1, ..., L\}$, $T^{(l)}$ is defined by (4.6).

4.1.2.2 Obtaining discrete analogues of Schwarz alternating algorithms

Suppose that we have only two subsets J_1 and J_2 and that $J_1 \bigcap J_2 \neq \emptyset$, so we have an overlapping between the 1st and the 2nd subdomains and

$$A = M_1 - N_1 = M_2 - N_2$$

Consider the matrices E_{lk} such that

$$(E_{11})_{i,i} = \begin{cases} 1 \ \forall i \in J_1 \\ 0 \ \forall i \notin J_1 \end{cases}, \quad (E_{12})_{i,i} = \begin{cases} 0 \ \forall i \in J_1 \\ 1 \ \forall i \notin J_1 \end{cases} \qquad (4.9)$$

$$(E_{21})_{i,i} = \begin{cases} 1 \ \forall i \notin J_2 \\ 0 \ \forall i \in J_2 \end{cases}, \quad (E_{22})_{i,i} = \begin{cases} 0 \ \forall i \notin J_2 \\ 1 \ \forall i \in J_2 \end{cases}$$

Define the fixed point mapping

$$T^S(x^1, x^2) = (y^1, y^2) \ such \ that \ for \ l = 1, 2$$

$$\begin{cases} y^l = T^{(l)}(z^l) \\ z^l = \sum_{k=1}^{2} E_{lk} x^k \end{cases} \qquad (4.10)$$

where for $l \in \{1, 2\}$, $T^{(l)}$ is defined by (4.6). Then the additive discrete analogue of the Schwarz alternating method corresponds to the successive approximation method applied to T^S, and the multiplicative discrete analogue of the Schwarz alternating method corresponds to the block nonlinear Gauss-Seidel method applied to T^S.

4.1.2.3 Obtaining discrete analogues of multisubdomain Schwarz algorithms

We introduce the weighting matrices E_k satisfying (4.8) and the matrices E_{lk} such that for $l \in \{1, ..., L\}$

$$(E_{ll})_{i,i} = \begin{cases} 1 \ if \ i \in J_l \\ 0 \ if \ i \notin J_l \end{cases}$$

$$(E_{lk})_{i,i} = \begin{cases} 0 \ if \ i \in J_l \\ (E_k)_{i,i} \ if \ i \notin J_l \end{cases} \qquad (4.11)$$

the synchronous iterations corresponding to the discrete analogue of the multisubdomain Schwarz method are defined by the fixed point mapping T^{MS}

$$T^{MS}(x^1, ..., x^L) = (y^1, ..., y^L) \ such \ that$$

$$\begin{cases} y^l = T^{(l)}(z^l) \\ z^l = \sum_{k=1}^{L} E_{lk} x^k \end{cases} \tag{4.12}$$

where E_{lk} are defined by (4.11) and $T^{(l)}$ are defined by (4.6).

4.1.2.4 Convergence of multisplitting and *two-stage* multisplitting algorithms

As mentioned above, the convergence of synchronous multisplitting algorithms was established by O'Leary and White in [92] when $A^{-1} \geq 0$, where A is the matrix of the linear system, and the splittings are weak regular.

When the linear systems, arising from the multisplitting algorithm, are not solved exactly but are instead approximated by iterative methods, we are confronted with *two-stage multisplitting algorithms* or *inner-outer iterations*. The convergence of two-stage multisplitting algorithms when the number of inner iterations is fixed was established by Szyld and Jones in [109] (see also [75]) for $A^{-1} \geq 0$ when the outer splittings are regular and the inner splittings are weak regular. The convergence of two-stage multisplitting algorithms was also studied by Bahi et al. [27] in a more general context including linear and nonlinear systems of equations.

4.2 Nonlinear systems: parallel synchronous Newton-multisplitting algorithms

Now we are interested in the development of parallel algorithms for nonlinear problems. We concentrate on the Newton method, since it is the most commonly used method to solve nonlinear systems.

4.2.1 Newton-Jacobi algorithms

Consider the nonlinear problem

$$F(x) = 0 \tag{4.13}$$

and the Newton method defined in Chapter 2 by the iterations

$$x^{(k+1)} = x^{(k)} - F'(x^{(k)})^{-1} F(x^{(k)}).$$

The solution of the system

$$F'(x^{(k)}) x^{(k+1)} = F'(x^{(k)}) x^{(k)} - F(x^{(k)}) \tag{4.14}$$

may be particularly prohibitive when the dimension of the problem is large. In this case, we can use an iterative method instead of direct ones in order to obtain an approximate solution of the system (4.14).

Let $D^{(k)}$ be a block diagonal matrix of $F'(x^{(k)})$, then

$$F'(x^{(k)}) = D^{(k)} - D^{(k)} + F'(x^{(k)}). \qquad (4.15)$$

The linear system (4.14) can be solved by the block Jacobi algorithm associated with the splitting (4.15). The obtained scheme consists in computing the iteration vectors by the following *two-stage* algorithm.

Algorithm 4.2 Newton-Jacobi scheme

Choose any arbitrary initial vector $(x^{(0)})^{(0)} = 0$
for k = 1,2,... do
   for l = 1,2,... do
      $D^{(k)}(x^{(k+1)})^{(l+1)} \leftarrow (D^{(k)} - F'(x^{(k)}))(x^{(k)})^{(l)} - F(x^{(k)})$
   end for
end for

It should be noticed that in practice only a fixed number of inner iterations is performed and that the number of inner iterations may vary in function of the Newton outer iterations. We are then in the presence of nonstationary iterative methods. The next section introduces Newton-multisplitting algorithms which are a generalization of Newton-Jacobi algorithms.

4.2.2   Newton-multisplitting algorithms

We suppose that (4.13) has a solution x^* and that F is Fréchet differentiable on a neighborhood of x^*. We also suppose that F' is nonsingular and Lipschitz continuous on a neighborhood of x^*. Newton iterations can be rewritten in the form

$$x^{(k+1)} = x^{(k)} - y^{(k)}, \ k = 0, 1, 2, ...$$

where $y^{(k)}$ is the solution of the linear system

$$F'(x^{(k)})y = F(x^{(k)}) \qquad (4.16)$$

Using an iterative method to solve (4.16) gives rise to the so-called *Newton iterative methods* [5], [6]. In [117], White proposes the parallel Newton-SOR method in order to solve nonlinear systems on parallel computers. In [5] and [6], the authors propose nonstationary multisplitting methods to solve (4.16), i.e., they consider for each k, a collection of L splittings of $F'(x^{(k)})$,

$$F'(x^{(k)}) = M_l(x^{(k)}) - N_l(x^{(k)}), \ l = 1, ..., L, \qquad (4.17)$$

Suppose that the weighting matrices (4.5) only depend on one index and that the solution of system (4.16) is approximated by performing q iterations of the multisplitting method and that $y^{(0)} = 0$.

The parallel Newton-multisplitting method can be written as follows:

$$x^{(k+1)} = G(x^{(k)}),\tag{4.18}$$

where

$$G(x) = x - A(x)F(x),$$

and

$$A(x) = \sum_{l=1}^{L} E_l(x)(I - (M_l(x)^{-1}N_l(x))^q(F'(x))^{-1}.\tag{4.19}$$

The following result gives the convergence condition of synchronous Newton-multisplitting algorithms.

THEOREM 4.4
If the splittings (4.17) are convergent, then there exists a neighborhood V_{x^*} of the solution x^*, such that any synchronous Newton-multisplitting algorithm associated with (4.18) and (4.19) starting from $x^{(0)} \in V_{x^*}$ converges to x^*.

PROOF   We have

$$G'(x^*) = I - A(x^*)F'(x^*).\tag{4.20}$$

From (4.19) we have

$$G'(x^*) = I - \sum_{l=1}^{L} E_l(x^*)(I - (M_l(x^*)^{-1}N_l(x^*))^q.\tag{4.21}$$

The properties of the weighting matrices imply that

$$G'(x^*) = \sum_{l=1}^{L} E_l(x^*)(M_l(x^*)^{-1}N_l(x^*))^q.\tag{4.22}$$

As the splittings (4.17) are convergent, we deduce by the application of proposition 3.2 of [27] that

$$\rho(G'(x^*)) \leq \max_{1 \leq l \leq L} \rho((M_l(x^*)^{-1}N_l(x^*))^q) < 1.$$

The result follows from Ostrowski theorem [95] (see the Appendix).   ☐

One can also suppose that the approximate solution of (4.16) is done by performing different $q_{k,l}$ inner linear iterations based on the linear splittings as explained in [5]. The obtained two-stage algorithm is called a *nonstationary Newton iterative algorithm*. The following convergence result is proved in [6].

THEOREM 4.5

If any of the following two conditions is satisfied, then there exists a neighborhood V_{x^*} such that the Newton-multisplitting started with $x^{(0)} \in V_{x^*}$ converges to x^*,

1. $F'(x^*)$ *is monotone and the splittings (4.17) are weak regular.*

2. $F'(x^*)$ *is an H-matrix and the splittings (4.17) are H-compatible (see the Appendix).*

4.3 Preconditioning

Some preconditioning algorithms have been adapted to parallel synchronous algorithms. In this case, depending on the amount of communications and of synchronizations, on the granularity of the method, on the degree of parallelism and especially on the network efficiency, the performances of those algorithms are relatively limited with a large number of processors. Nevertheless, some preconditioners have been designed for parallel architectures. For example, parallel preconditioners, based on ILU, are very sensitive to the ordering of the unknowns. The more independent the unknowns are, the more efficient the parallel preconditioner ILU is. For more explanations on parallel preconditioners, interested readers are invited to read [29, 102, 121, 41] and the references therein.

4.4 Implementation

Implementing a synchronous parallel algorithm depends on the platform used to execute it. In fact, it is possible to distinguish at least two different paradigms from the programming point of view. The first one is only dedicated to shared memory architectures. The second one is commonly called *message passing* and is mainly used in distributed architectures.

On shared memory architectures, at least two different kinds of programming exist. The first class aims at parallelizing most consuming loops. Consequently we obtain what is usually called a *data parallel* code which fits the class of fine grained parallelism. In this model, the same set of instructions runs simultaneously on different pieces of data. More precisely, each processor executes only a part of the loop. The other parts are achieved by other processors. In this context, the best-known programming model is certainly OpenMP [36] for which a programmer only needs to add compiler directives in its code. In some particular cases the programmer does not need to modify its code. Compiler directives are interpreted in order to split the initial work of loops into smaller parts that are executed by the available processors.

The second class aims at decomposing the work into large parts with smaller parts in which processors exchange some data. Usually this method is efficient if the parts are relatively independent. This model is frequently called *coarse grained* parallelism.

In both these models, processors can access to the whole memory for reading and writing. If two processors access the same data in writing, the behavior is often nondeterministic. So the programmer must carefully check that only one processor writes into a part of data at each instant. However, the big advantage of this model is its programming simplicity since a processor can directly read any part of the memory without asking it to any processor. Of course according to the architecture, the time to read data is not always constant and this often leads to bottlenecks.

Concerning the programming of synchronous iterative algorithms both models are interesting but do not provide the same programming work. Using fine grained parallelism with loops splitting is quite easy. Nevertheless such codes are not as scalable as coarse grained parallelism codes which require a longer programming endeavor. Now, most programmers do agree with the fact that the transformation of a sequential program into a parallel one using shared memory mechanism requires less work than using other parallelization paradigms. Moreover, as few architectures provide a shared memory mechanism, an application parallelized using that paradigm would not be as portable as if it were parallelized using another programming model.

An interesting alternative, if we are interested in code reuse, lies in designing an application with the message passing paradigm. This is the classical model used for distributed architectures in which processors communicate by sending/receiving messages to/from each other. Using a message passing paradigm often requires rather a lot more work and time to design a parallel application compared to using a shared memory paradigm. Nonetheless, such a program is more portable since it can be executed on many architectures: either distributed ones or shared memory ones. Even if it is not as efficient on shared memory architecture as using a shared memory paradigm, it is possible to run a program designed with a message passing paradigm on such an architecture. Message passing programs generally require the use of buffers in order to send or receive messages; that is why on shared memory architectures, they could be less efficient.

With the development of multi-core processors, scientists have access to clusters in which both paradigms can be used in order to benefit from the best performance. On the one hand, communications between processors linked by a network should be achieved using a message passing interface. On the other hand, inside a multiprocessor or a multi-core machine, a shared memory paradigm is preferable to obtain efficient codes. So, in this kind of architecture, which will probably be used more and more in the next years, the mixing of shared memory and message passing paradigms will probably be the best solution in order to obtain efficient codes.

Designing a parallel algorithm to solve a linear system or a nonlinear one requires approximately the same notions from the programming point of view. First it is extremely important to be rigorous. Of course this is important for implementing any sequential algorithm but it is more important as far as parallel application is concerned, the least error being indeed extremely difficult to detect. Implementing a parallel algorithm depends on the parallel language chosen. Nevertheless, with some experience, it is relatively easy to disregard it. This is quite similar to implementing a sequential algorithm with one language or with another one which only differs by the syntax. That is why in the following we will only focus on a general implementation that will slightly differ from using MPI with the language C or OpenMP with C++ and a coarse-grained paradigm.

4.4.1   Survey of synchronous algorithms with shared memory architecture

As shared memory architectures are not as scalable as distributed ones, in this book, we only describe the principles for designing parallel synchronous iterative algorithms using this kind of architecture. Moreover, parallelizing an application using a shared memory architecture is often easier than using a distributed one.

Since the parallelization of an iterative algorithm using the coarse grained paradigm with a shared memory architecture is quite similar to the parallelization of the same algorithm with a distributed system, we only focus on the fine grained paradigm. In addition, this paradigm is easier to implement and is probably the most used model with a limited number of processors, as it is the case with cheaper architectures (small multi-processor or multi-core systems).

Since OpenMP is the most used tool to build fine grained parallel algorithms, we limit our explanation to it. Roughly speaking, iterative algorithms are all composed of a loop which represents an iteration. As an iteration uses the results computed in previous iterations it is not possible to execute different iterations concurrently on different processors. Therefore, it is necessary to parallelize the computation inside an iteration. Consequently, the number of iterations with such a parallel program will be exactly the same as the sequential code used for the parallelization. The fact of using the fine grained paradigm with OpenMP consists in parallelizing all the loops inside an iteration as soon as the work inside a loop can be done concurrently. It should be noticed that several loops can be parallelized inside an iterative algorithm. For example, scalar products, matrix-vector products, or vector additions can be executed using parallelization at the loop level. Taking the Jacobi algorithm (Algorithm 2.1), it is possible to parallelize the first loop indexed by i. Concerning the Gauss-Seidel algorithm (Algorithm 2.2), it is well-known that this algorithm is less parallelizable. It is not possible to parallelize the same loop. Nevertheless it is possible to parallelize the inner loops, indexed by j in that algorithm. Of course, this parallel scheme is less efficient since the parallelization of small loops only provides a small performance gain. The parallelization of nonlinear methods is sensibly similar. Only some loops inside an iteration are parallelizable. Practically speaking, this kind of parallelization provides good speed-ups with few processors, but unfortunately, bad speed-ups as soon as the number of processors increases.

Coarse grained parallelism implementation is completely different and uses approximately the same scheme either with a shared memory architecture or with a distributed one (except for the communication handling). In the following sections we distinguish the case of sequential algorithms that can be parallelized using traditional message passing schemes from the Jacobi algorithm and multisplitting ones for which an appropriate implementation will enable them to be executed in an asynchronous mode as we will see in Chapter 5.

4.4.2 Synchronous Jacobi algorithm

It is well known that it is generally difficult to implement an efficient and general code. According to the structure of the studied matrix, some optimizations can produce very efficient codes. In the following we give an implementation of the Jacobi parallel code. In order to keep this algorithm simple, we consider that a processor needs to send its results to all other processors. In many cases, it is easy to compute the list of neighbors for each processor and, consequently, only send the results to processors that require it. So, we consider that the matrix A is split into rectangular parts as in Figure 4.3. The vector B is split as the vector X.

Each processor only owns a part of the vector X but it requires a larger part of the vector $XOld$. According to the structure of the matrix A, this part may vary. If A is an almost dense matrix, then each processor needs the totality of the vector $XOld$. Algorithm 4.3 illustrates the synchronous Jacobi algorithm. In this algorithm $Size$ and $SizeGlo$ represent, respectively, the local size and the global size of the matrix A, $Offset$ represents the offset of the global index for the computation, i.e., the sum of the local size of all parts of matrices of processors having a lower rank. The variable $MyRank$ represents the rank of the processor, that is to say the number of the current processor in the computation. In order to detect the convergence, we first compute at each iteration the local convergence and put the result in the variable $Error$. Then we use a reduce operation that computes the maximum of all the local errors. With MPI, such an operation is directly implemented in the API.

Without considering the communications, this parallel version of the Jacobi method is completely similar to the sequential one. In the algorithm we are using a high level communication procedure called $AllReduce$. The goal of this procedure consists in applying a reduction operation on the variable $Error$

Algorithm 4.3 Synchronous Jacobi algorithm

NbProcs: number of processors
MyRank: rank of the processor
Size: local size of the matrix
SizeGlo: global size of the matrix
Offset: offset of the global index
A[Size][SizeGlo]: local part of the matrix
X[Size]: local part of the solution vector
XOld[SizeGlo]: global solution vector
B[Size]: local part of the right-hand side vector
Error: local error
MaxError: global error
Epsilon: desired accuracy

repeat
  for i=0 to Size−1 **do**
    X[i] ← 0
    for j=0 to Offset−1 **do**
      X[i] ← X[i]+A[i][j]×XOld[j]
    end for
    for j=Offset+Size to SizeGlo−1 **do**
      X[i] ← X[i]+A[i][j]×XOld[j]
    end for
  end for
  for i=0 to Size−1 **do**
    X[i] ← (B[i]−X[i])/A[i][i+Offset]
  end for
  Error← 0
  for i=0 to Size−1 **do**
    Error ← max(Error, abs(X[i]−XOld[i+Offset]))
    XOld[i+Offset] ← X[i]
  end for
  for k=0 to NbProcs−1 **do**
    if k ≠ MyRank **then**
      Send(k, X)
    end if
  end for
  for k=0 to NbProcs−1 **do**
    if k ≠ MyRank **then**
      Recv(k, XOld[k×Size])
    end if
  end for
  AllReduce(Error, ErrorMax, MAX)
until stopping criteria is reached (MaxError ≤ Epsilon)

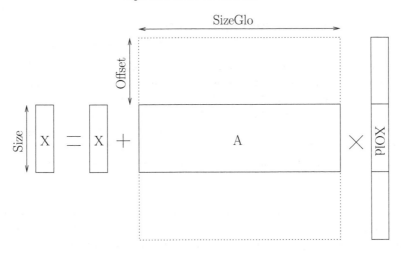

FIGURE 4.3: Splitting of the matrix for the synchronous Jacobi method.

and putting the result in the variable *ErrorMax*. In that case, the MAX operator is used in order to compute the maximum of the local error (or norm here). The result is available on each processor. As a consequence, the result is strictly identical to the result obtained using the sequential version (neglecting the potential rounding errors).

The communication part consists for each processor in sending its results (the vector X) to all the processors that need it. In the case of a dense matrix, processor k sends its result to all the other processors. Then a processor receives all the results and directly puts them in their right places in the array *XOld*; each place is computed in function of the rank of the sender and the (local) size of the matrix (considered constant). From a practical point of view, the programmer should rather use nonblocking communications in order to perform the exchange of data. According to the programming environment, it is possible that the use of blocking sends and receptions leads to a deadlock situation in which some processors are blocked while communicating simultaneously with each other.

With some high level implementation libraries (like MPI) it is possible to use a single call to a function to realize this exchange operation. In the code corresponding to this example we use it, so interested readers are invited to test it with MPI. For example, it is possible to use a procedure called *AllToAllV(X[Offset],X,Size)* that produces the same results as the communication part in Algorithm 4.3.

4.4.3   Synchronous conjugate gradient algorithm

The conjugate gradient algorithm can be parallelized using basically the same procedure, i.e., splitting the work and synchronizing every part of the code requiring it. Algorithm 4.4 describes the synchronous version of the conjugate gradient algorithm. In the following, we use *X[Offset]=Y* in order to copy the elements of the vector *Y* into *X* at the offset *Offset*. Likewise, for some operations we specify the number of elements that are concerned. For example, *P[Offset]=R (copy Size elements)* means that size elements of *R* are copied into *P* with the offset *Offset*.

Compared to the synchronous parallel Jacobi algorithm, the parallel version of the conjugate gradient requires more synchronizations. With the parallel Jacobi only two steps act as a synchronization, the exchange of data and the computation of the global error. In opposition, the parallel conjugate gradient algorithm contains twice the synchronization steps: three *AllReduce* operations and one *AllToAllV*.

4.4.4   Synchronous block Jacobi algorithm

The synchronous block Jacobi algorithm is relatively easy to write having the sequential version in mind. In fact, each processor is responsible for the computation of a block and after an iteration, all processors send their local solution to their neighbors that need it. In Algorithm 4.5, we describe the synchronous version of this algorithm. At the end of an iteration, processors exchange their local computation using an *AllToALLV* procedure. Then, they compute the global error. Consequently this algorithm requires two synchronization steps. In order to solve the local subsystem, each processor uses an appropriate method. In practice, depending on the size of the submatrix and its degree of density, a sparse or a dense direct method can be used. The solution of a subsystem is considered exact (neglecting rounding errors), so an iterative method is not considered for solving a local subsystem.

At each iteration, three main steps may be distinguished in the block Jacobi algorithm. The first one consists in updating the right-hand side using the dependencies of other processors. In Algorithm 4.5, this step updates the vector *BTmp*. The second step aims at solving the local subsystem on each processor. Using an existing solver obviously simplifies the programming of this method. According to the characteristics of the matrix, the choice of the inner solver may drastically change the efficiency of the parallel solver. Finally, the third step corresponds to the data exchanges and to the global error computation.

Algorithm 4.4 Synchronous conjugate gradient algorithm

NbProcs: number of processors
MyRank: rank of the processor
Size: local size of the matrix
SizeGlo: global size of the matrix
Offset: offset of the global index
A[Size][SizeGlo]: local part of the matrix
X[SizeGlo]: solution vector
R[Size]: local part of the residual vector
B[Size]: local part of the right-hand side vector
P[SizeGlo]: search direction vector
Q[Size]: local part of the orthogonal vector to the search direction
DotPQ, DotPQGlo: local and global scalar product of (P,Q)
Alpha, Beta, Rho, RhoGlo: scalar variables
Error: local error
MaxError: global error
Epsilon: desired accuracy

Offset $\leftarrow$ Size$\times$MyRank
R $\leftarrow$ B$-$A$\times$X
repeat
   Rho $\leftarrow$ (R,R)
   AllReduce(Rho, RhoGlo, SUM)
   if i=1 **then**
      P[Offset] $\leftarrow$ R *(copy Size elements)*
   else
      Beta $\leftarrow$ RhoGlo/RhoOldGlo
      P[Offset] $\leftarrow$ R+Beta$\times$P[Offset] *(copy Size elements)*
   end if
   AllToAllV(P[Offset], P, Size)
   Q $\leftarrow$ A$\times$P
   DotPQ $\leftarrow$ (P[Offset],Q) *(for only Size elements)*
   AllReduce(DotPQ, DotPQGlo, SUM)
   Alpha $\leftarrow$ RhoGlo/DotPQGlo
   X[Offset] $\leftarrow$ X+Alpha$\times$P[Offset] *(for Size elements)*
   R $\leftarrow$ R$-$Alpha$\times$Q
   RhoOldGlo $\leftarrow$ RhoGlo
   Error $\leftarrow$ 0
   for i=0 to Size$-$1 **do**
      Error $\leftarrow$ max(Error, abs(R[i]))
   end for
   AllReduce(Error, ErrorMax, MAX)
until stopping criteria is reached (MaxError $\leq$ Epsilon)

Algorithm 4.5 Synchronous block Jacobi algorithm

NbProcs: number of processors
MyRank: rank of the processor
Size: local size of the matrix
SizeGlo: global size of the matrix
Offset: offset of the global index
A[Size][SizeGlo]: local part of the matrix
X[Size]: local part of the solution vector
B[Size]: local part of the right-hand side vector
BTmp[Size]: intermediate local part of the right-hand side vector
XOld[SizeGlo]: global solution vector
Error: local error
MaxError: global error
Epsilon: desired accuracy

Offset← Size×MyRank
repeat
   for i=0 to Size−1 **do**
     BTmp[i]← B[i]
   end for
   for i=0 to Size−1 **do**
     for j=0 to Offset−1 **do**
       BTmp[i]← BTmp[i]−A[i][j]×XOld[j]
     end for
     for j=Offset+Size to SizeGlo **do**
       BTmp[i]← BTmp[i]−A[i][j]×XOld[j]
     end for
   end for
   X← Solve(A, BTmp)
   Error← 0
   for i=0 to Size−1 **do**
     Error ← max(Error, abs(X[i]−XOld[i+Offset]))
     XOld[i+Offset]← X[i]
   end for
   AllToAllV(XOld[Offset], XOld, Size)
   AllReduce(Error, ErrorMax, MAX)
until stopping criteria is reached (MaxError ≤ Epsilon)

4.4.5 Synchronous multisplitting algorithm for solving linear systems

This algorithm comes directly from the formulation of Equations (4.3)-(4.5). In the following we explain how to obtain the algorithm without overlapping components and then we introduce the overlapping of components.

The first step consists in defining the weighting matrices. Without introducing overlapping, those matrices are diagonal and either contain 1 or 0 on the diagonal. For example, if we take three processors ($L = 3$), the weighting matrices are defined as in Figure 4.4.

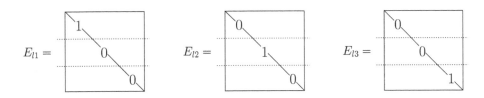

FIGURE 4.4: An example with three weighting matrices.

Having defined those matrices, we need to define the matrices M_l and N_l. With Equation (4.4) and by defining $T^{(l)}$ as in Equation (4.6), with three processors, we obtain the following system

$$\begin{cases} y^1 = M_1^{-1} N_1 (E_{11} x^1 + E_{12} x^2 + E_{13} x^3) + M_1^{-1} b \\ y^2 = M_2^{-1} N_2 (E_{21} x^1 + E_{22} x^2 + E_{23} x^3) + M_2^{-1} b \\ y^3 = M_3^{-1} N_3 (E_{31} x^1 + E_{32} x^2 + E_{33} x^3) + M_3^{-1} b \end{cases} \tag{4.23}$$

In that system, matrices M_l^{-1} and N_l and vectors y^l, x^l and b are not decomposed. From a practical point of view, a processor does not handle the whole vectors and matrices, it only has the parts it is in charge of. In the example with three processors, each processor has a third of data. So we can define y'^l, x'^l and b'^l that correspond to the parts handled by the processors. And Equation (4.23) can be rewritten as

$$\begin{cases} y'^1 = (A_{11}^{-1} \times -A_{12}) x'^2 + (A_{11}^{-1} \times -A_{13}) x'^3 + A_{11}^{-1} b'^1 \\ y'^2 = (A_{22}^{-1} \times -A_{21}) x'^1 + (A_{22}^{-1} \times -A_{23}) x'^3 + A_{22}^{-1} b'^2 \\ y'^3 = (A_{33}^{-1} \times -A_{31}) x'^1 + (A_{33}^{-1} \times -A_{32}) x'^2 + A_{33}^{-1} b'^3 \end{cases} \tag{4.24}$$

with, for example, the splittings depicted in Figure 4.5.

By multiplying each previous equation by A_{ii} we obtain:

$$\begin{cases} A_{11} \times y'^1 = b'^1 - A_{12} x'^2 - A_{13} x'^3 \\ A_{22} \times y'^2 = b'^2 - A_{21} x'^1 - A_{23} x'^3 \\ A_{33} \times y'^3 = b'^3 - A_{31} x'^1 - A_{32} x'^2 \end{cases} \tag{4.25}$$

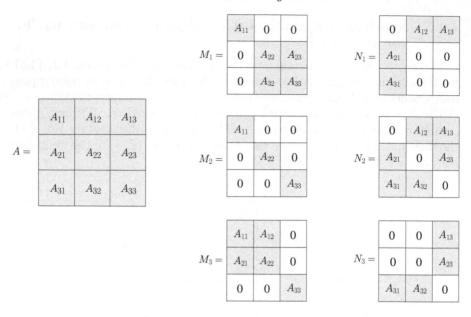

FIGURE 4.5: An example of possible splittings with three processors.

In this example, each processor i has a local linear subsystem to solve. The right-hand side contains the corresponding part of the vector b'^i minus all the block off-diagonal (A_{ij} with $j \neq i$) multiplied by the corresponding x'^j with $j \neq i$.

From a practical point of view, we call the off-diagonal block the dependencies of the computation. So, processor 1 depends, respectively, on processors 2 and 3 if blocks A_{12} and A_{13} are, respectively, nonempty. In the following practical algorithm we want to gather all dependencies before the current processor in an array called *LeftDep* and all the dependencies after the current processor in an array called *RightDep*. Likewise, each processor has two vectors *XLeft* and *XRight*.

In Figure 4.6, the decomposition of the matrix is illustrated. Each processor is in charge of a rectangular part of the matrix. This rectangular part is split into three parts. The left dependencies (DepLeft) involve components computed by processors whose rank is strictly smaller than the one of the considered processor. The submatrix (noted A) is the square matrix that a processor is in charge of; it corresponds to the matrix A_{ii} in Equation (4.25). And finally, the right dependencies (DepRight) involve components computed by processors whose rank is strictly greater than the one of the considered processor. With such a decomposition, a processor needs to solve:

$$A \times X = B - DepLeft \times XLeft - DepRight \times XRight \qquad (4.26)$$

which exactly corresponds to the lines in Equation (4.25). As soon as it has computed the solution of the subsystem, this solution needs to be sent to all processors depending on it.

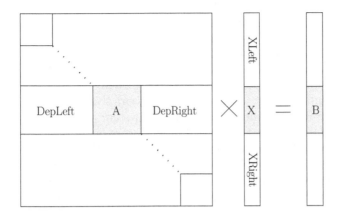

FIGURE 4.6: Decomposition of the matrix.

Algorithm 4.6 illustrates the synchronous version of the multisplitting algorithm to solve linear systems. At the beginning of an iteration, a processor computes the right-hand side as in Equation (4.26). Then it solves the linear system composed of the submatrix it is in charge of with the right-hand side that takes into account the dependencies just computed. To solve this linear system, it is possible to use either a direct algorithm or an iterative one (with or without using a preconditioner); in this latter case we obtain a two-stage method. In Section 6.4, we report an experiment that highlights the impact of using this or that algorithm. The next step consists in exchanging dependencies. Before starting the iterative process, processors exchange their dependencies in order to initialize the arrays $DependsOnMe$ and $IDependOn$. This part involves the use of the offset computed on each processor. Moreover, a processor takes into consideration what parts of the solution vector it needs. With those arrays, a processor knows its neighbors. Consequently, it can send to each of its neighbors the part of its vector that they need. Then it is ready to receive dependencies from its neighbors. According to the rank of a neighbor, a processor integrates the dependency in $DepLeft$ or $DepRight$. Globally, this exchange part acts as a synchronization step. The last step lies in computing the error locally and then globally using for a second time a synchronization step so that all the processors know the global error.

As explained, one particularity of the multisplitting method is that it allows the processors to overlap components in order to speed up the convergence. The principle is to let some processors compute simultaneously some com-

---

Algorithm 4.6 Synchronous linear multisplitting algorithm

NbProcs: number of processors
MyRank: rank of the processor
Size: local size of the matrix
SizeGlo: global size of the matrix
Offset: offset of the global index
A[Size][Size]: local block-diagonal part of the matrix
DepLeft[Size][Offset]: submatrix with left dependencies
DepRight[Size][SizeGlo-Offset-Size]: submatrix with right dependencies
DependsOnMe[NbProcs]: array of the dependent processors
IDependOn[NbProcs]: array of the processors this processor depends on
B[Size]: right-hand side vector of the subsystem
X[Size], XOld[Size]: local part solution vectors of the subsystem
XLeft[Offset]: left part of the solution vector of the system
XRight[SizeGlo-Offset-Size]: right part of the solution vector of the system
BLoc[Size]: array containing the local computations on the right-hand side
TLoc[Size]: array used for the receptions of the dependencies
Error: local error
MaxError: global error
Epsilon: desired accuracy

repeat
  BLoc ← B
  if MyRank≠0 **then**
    BLoc ← BLoc−DepLeft×XLeft
  end if
  if MyRank ≠ NbProcs−1 **then**
    BLoc ← BLoc−DepRight×XRight
  end if
  X ← Solve(A, BLoc)
  for i=0 to NbProcs−1 **do**
    if i ≠ MyRank and DependsOnMe[i] **then**
      Send(i, PartOf(X, i))
    end if
  end for
  for i=0 to NbProcs−1 **do**
    if i ≠ MyRank and IDependOn[i] **then**
      Recv(i, TLoc)
      Update XLeft or Xright with TLoc according to the processor *i*
    end if
  end for
  Error← 0
  for i=0 to Size−1 **do**
    Error ← max(Error, abs(X[i]−XOld[i]))
    XOld[i]← X[i]
  end for
  AllReduce(Error, ErrorMax, MAX)
until stopping criteria is reached (MaxError ≤ Epsilon)

---

ponents and to mix the results in order to obtain an accurate result faster. It corresponds to splitting the matrix into rectangular matrices that are not disjoint (J sets in the theoretical framework). In Figure 4.7, we give an example with a small matrix of size 9×9 for which the corresponding linear system is solved with three processors using one overlapped component with each neighbor. Without using overlapping, each processor has three components (processor 1 has components 1 to 3, processor 2 has components 4 to 6 and processor 3 has components 7 to 9). If we allow some components to be overlapped, processors with only one neighbor (i.e., processors 1 and 3 in the figure) have four components. Processor 2 has five components. In Figure 4.7, the hatched parts represent the submatrices that processors are in charge of (matrix A in Algorithm 4.6) and black dots represent non-null values of the matrix. Parts in the matrix that are doubly hatched highlight components that are computed by two processors. Subvectors x'^i (X in Algorithm 4.6) also have overlapped components that are represented in gray in the figure. Non-null values that are not in submatrices represent dependencies. In Figure 4.7, circled dots illustrate dependencies that are simultaneously computed by two processors. Dependencies on lines 1 and 2 are computed by processors 2 and 3. Likewise, dependencies on lines 8 and 9 are computed by processors 1 and 2.

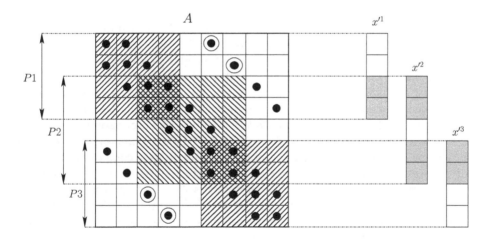

FIGURE 4.7: An example of decomposition of a 9×9 matrix with three processors and one component overlapped at each boundary on each processor.

There are multiple ways to mix the overlapped components. In the following, we explain four ways to mix overlapped components.

4.4.5.1   Overlapping strategy that uses locally computed values

In this case, a processor only uses components that it has computed while ignoring the values corresponding to the same set of components computed by its neighbors. With the example defined in Figure 4.7, it consists in defining the weighting matrices as in Figure 4.8. In this figure, matrices E_{1k}, E_{2k} and E_{3k}, respectively, correspond to weighting matrices of processors 1, 2 and 3. Overlapped components are represented in gray in the figure. For example, components 3 and 4 are computed simultaneously by processors 1 and 2. For a processor i, we can remark that the sum of weighted matrices E_{ik} is equal to the identity matrix (as expressed in Equation (4.5)). Hence, with this strategy, processor 1 uses its components 1 to 4, it uses components 5 and 6 of processor 2 and components 7 to 9 of processor 3. Processor 2 uses its components 3 to 7 and uses components 1 and 2 of processor 1 and components 8 and 9 of processor 3. Processor 3 proceeds similarly to processor 1; since it has its 4 components, it uses 2 components of processor 2 and 3 components of processor 1. This strategy is quite easy to implement since it does not require any mixing of overlapped components. It simply consists in using all the components computed by a processor.

4.4.5.2   Overlapping strategy that uses values computed by close neighbors

This strategy has similarities to the previous one because it does not require any mixing of overlapped components either. The principle consists in using all the overlapped components of its close neighbors. Weighting matrices, corresponding to the same example, are illustrated in Figure 4.9. In that case, processor 1 is close to processor 2 but not to processor 3. So processor 1 only uses its components 1 and 2, it uses components 3 to 6 that are computed by processor 2 and it uses components 7 to 9 computed by processor 3. Processor 2 has two close neighbors (processors 1 and 3); it uses components 1 to 4 of processor 1, it uses its single component 5 and it uses components 6 to 9 of processor 3. Processor 3 proceeds similarly to processor 1 and it uses components 1 to 3 of processor 1, components 4 to 7 of processor 2 and its components 8 and 9. Compared to the previous strategy, this one requires more data exchange since a processor requires all overlapped components of its close neighbors.

4.4.5.3   Overlapping strategy that mixes overlapped components with close neighbors

With this strategy a processor mixes its overlapped components with its close neighbors. In Figure 4.10, we give an example of the mixing which consists in taking half of the value computed by a processor and half of the value computed by the close neighbor. So, processor 1 has its components 1 and 2 and it mixes its components 3 and 4 with processor 2, it uses components

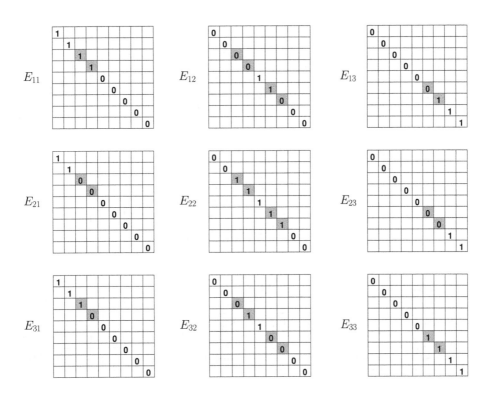

FIGURE 4.8: Overlapping strategy that uses values computed locally.

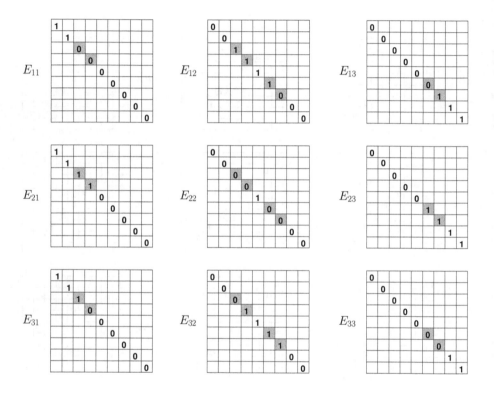

FIGURE 4.9: Overlapping strategy that uses values computed by close neighbors.

5 and 6 of processor 2 and components 7 to 9 of processor 3. Processor 2 uses components 1 and 2 of processor 1, it mixes its components 3 and 4 with processor 1, it uses it component 5, it mixes its components 6 and 7 with processor 3 and it uses components 8 and 9 of processor 3. Processor 3 uses components 1 to 3 of processor 1, it uses components 4 and 5 of processor 2, it mixes its components 6 and 7 with processor 2 and it uses its components 8 and 9. Compared to the previous strategy, the amount of data exchange is strictly equal. With this strategy it is possible to use different ratios of mixing as soon as the sum of ratios on one line of all the matrices E_{lk} is equal to 1. For instance, it is possible to take 75% of the computed components and 25% of the values computed by the other processor.

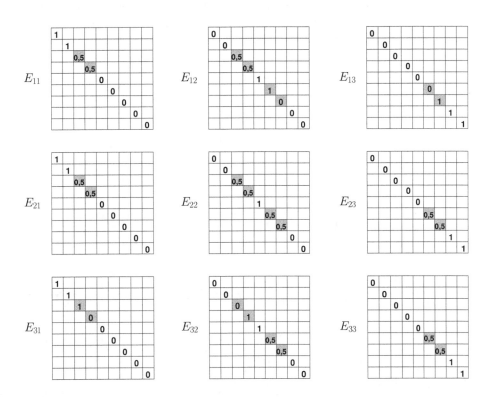

FIGURE 4.10: Overlapping strategy that mixes overlapped components with close neighbors.

4.4.5.4 Overlapping strategy that mixes all overlapped components

This strategy mixes all components that are overlapped (not only with its close neighbors). In Figure 4.11 we illustrate a possible example of values of the weighted matrices E_{lk} for this strategy. All the gray parts, corresponding to overlapped components, contain values that are different from 0 and 1. This strategy offers the most freedom to mix overlapping components. In our example, all processors use components 1 and 2 from processor 1, they mix components 3 and 4 from processors 1 and 2, they use component 5 of processor 2, they mix components 6 and 7 from processors 2 and 3 and they use components 8 and 9 of processor 3. If different values of mixing are used according to processors, all overlapped components must be sent to processors that need them. Consequently the amount of data transfered may be greater.

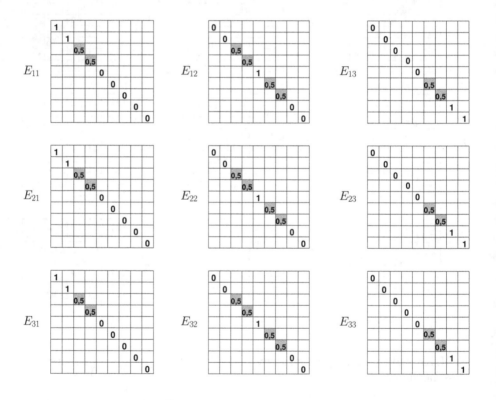

FIGURE 4.11: Overlapping strategy that mixes all overlapped values.

Implementing all those strategies in Algorithm 4.6 is quite easy. The matrix distribution should be achieved by taking into account the overlapped components on each processor. So, the offsets and the local size of matrices are different. Then, according to the overlapping strategy chosen, the amount of data might be different. For example with the first strategy presented, the amount of data is smaller, whereas with the last strategy the amount of data transfered is greater. According to the structure and dependencies of the problem, the choice of the best strategy is not a trivial task.

4.4.6 Synchronous Newton-multisplitting algorithm

In the previous sections we have explained how to implement the multisplitting method in order to solve linear systems. Now we describe how to design the algorithm based on the Newton-multisplitting method. As described formerly, this method uses the Newton method to linearize the problem. Using synchronous iterations, the algorithm first builds the Jacobian matrix and then it needs to solve the linear system obtained. The multisplitting method to solve a linear system is used for this. In Algorithms 4.7 and 4.8, both methods are coupled. The outer iterations perform the Newton iteration, whereas the inner ones solve the linear system. Variables of this algorithm are described in Algorithm 4.7 and its core is given in Algorithm 4.8.

At each Newton iteration, a processor starts by computing the rectangular part of the Jacobian matrix it is in charge of. Then it computes the right-hand side of the nonlinear function as described in Equation (4.16). Of course, each processor has a different set of components according to the block distribution. As soon as the Jacobian submatrices have been defined simultaneously on processors, they start to solve the linear system using the multisplitting algorithm for linear systems. To solve the subsystem it is possible to use a direct method or an iterative one. In this latter case, we obtain a two-stage algorithm. The Jacobian is computed using the same computation as in the sequential algorithm. The only difference is that the Jacobian is distributed on all processors, as described in Figure 4.12. The computation of $-F$ is also distributed across the processors.

After the computation of the Jacobian, the method consists in solving the whole linear system using the multisplitting method for linear systems. So, until global convergence of the multisplitting algorithm, a processor computes the local right-hand side and updates it using the dependencies computed by its neighbors. In the algorithm, J represents the local submatrix for a processor. $JDepLeft$ and $JDepRight$, respectively, correspond to $DepLeft$ and $DepRight$ defined in the multisplitting algorithm for linear systems, described previously. At each multisplitting iteration, a processor solves the subsystem composed of the Jacobian submatrix and the right-hand side using a sequential solver. Then, it sends its DX vector part to each of its neighbors that need it, and it receives the part of the solution of its neighbors and updates the vectors $DXLeft$ and $DXRight$. Finally, it computes the local error of the multisplitting process. When the multisplitting method has globally converged, a processor computes the local error of the Newton process. Pro-

Algorithm 4.7 Variables used in the synchronous Newton-multisplitting algorithm

NbProcs: number of processors
MyRank: rank of the processor
Size: local size of the matrix
SizeGlo: global size of the matrix
Offset: offset of the global index
J[Size][Size]: local block-diagonal part of the Jacobian matrix
JDepLeft[Size][Offset]: submatrix with left dependencies of the Jacobian
JDepRight[Size][SizeGlo-Offset-Size]: submatrix with right dependencies of the Jacobian
DependsOnMe[NbProcs]: array of the dependent processors
IDependOn[NbProcs]: array of the processors this processor depends on
F[Size], FLoc[Size]: right-hand side vectors of the subsystem
X[SizeGlo]: solution vector of the Newton subsystem
DX[Size], DXOld[Size]: local part solution vectors of the multisplitting subsystem
DXLeft[Offset]: left part of the solution vector of the system
DXRight[SizeGlo-Offset-Size]: right part of the solution vector of the system
TLoc[Size]: array used for the receptions of the dependencies
ErrorNewton: local error of the Newton process
MaxErrorNewton: global error of the Newton process
ErrorMulti: local error of the multisplitting
MaxErrorMulti: global error of the multisplitting process
EpsilonMulti: desired accuracy for the multisplitting process
EpsilonNewton: desired accuracy for the Newton process

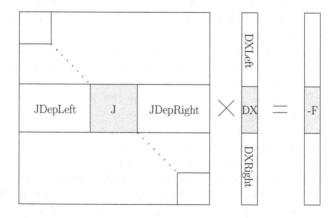

FIGURE 4.12: Decomposition of the Newton-multisplitting.

Algorithm 4.8 Synchronous Newton-multisplitting algorithm

---

repeat

    if first iteration or required **then**

      Computation of the Jacobian rectangular matrix and storage of the respective parts into J, $JDepLeft$ and $JDepRight$

    end if

    Computation of $-F$ depending on X from components Offset to Offset+size-1 and storage of the result into F

    repeat

      FLoc $\leftarrow$ F

      if MyRank $\neq 0$ **then**

        FLoc $\leftarrow$ FLoc$-$JDepLeft$\times$DXLeft

      end if

      if MyRank $\neq$ NbProcs-1 **then**

        FLoc $\leftarrow$ FLoc$-$JDepRight$\times$DXRight

      end if

      DX $\leftarrow$ Solve(J, FLoc)

      for i=0 to NbProcs-1 **do**

        if i $\neq$ MyRank and DependsOnMe[i] **then**

          Send(i, PartOf(DX, i))

        end if

      end for

      for i=0 to NbProcs-1 **do**

        if i $\neq$ MyRank and IDependOn[i] **then**

          Recv(i, TLoc)

          Update DXLeft or DXRight with TLoc according to processor i

        end if

      end for

      ErrorMulti$\leftarrow 0$

      for i=0 to Size-1 **do**

        ErrorMulti $\leftarrow$ max(ErrorMulti, abs(DX[i]$-$DXOld[i]))

        DXOld[i]$\leftarrow$ DX[i]

      end for

      AllReduce(ErrorMulti, MaxErrorMulti, MAX)

    until stopping criteria of multisplitting is reached

        (MaxErrorMulti $\leq$ EpsilonMulti)

    ErrorNewton$\leftarrow 0$

    for i=0 to Size-1 **do**

      X[Offset+i] $\leftarrow$ X[Offset+i]+DX[i]

      ErrorNewton $\leftarrow$ max(ErrorNewton, abs(DX[i]))

    end for

    AllToAllV(X[Offset], X, Size)

    AllReduce(ErrorNewton, MaxErrorNewton, MAX)

until stopping criteria of Newton is reached

      (MaxErrorNewton $\leq$ EpsilonNewton)

---

cessors execute Newton iterations until the global error is lower than a given threshold.

Analyzing the number of synchronizations of this algorithm we can remark that there are as many synchronizations as multisplitting iterations and Newton iterations (considering that there is only one synchronization per iteration, we have seen previously that this is not the case). Thus this algorithm requires quite an important number of synchronizations.

4.5   Convergence detection

We discuss here the problem of convergence detection in iterative processes of the form:

$$x^{(k+1)} = G(x^{(k)}) \tag{4.27}$$

where $x^{(k+1)}$ and $x^{(k)}$ are the global state vectors at the respective iterations $k+1$ and k, and G is a contraction. The useful property of contractions is that their convergence is ensured. This is why most of the iterative algorithms currently in use are contractions. In fact, an important constraint when designing an iterative method is precisely to ensure that it is a contraction.

However, most of the theoretical results related to the convergence of iterative processes, including contractions, are of limited interest in practice since they are often based on properties which are not directly calculable or whose computation cost is of the same order as the problem to solve. Hence, as explained in [33], practical methods for proving the convergence of iterative methods generally consist in finding a suitable norm for which it can be shown that each iteration reduces the distance between the current global state vector and the fixed point which represents the solution of the problem.

Unfortunately, it is possible to ensure that an iterative process is a contraction without being able to find the suitable norm which allows us to detect its convergence in practice. For example, in linear problems, we know that we have a contraction when the spectral radius of the iteration matrix is smaller than one but there is no information about the norm to use in practice. Moreover, as already mentioned in Section 1.2, a contraction is norm dependent and may be contractive with a given norm and non-contractive with another one. This is an important problem which may induce some difficulties in the convergence detection.

Indeed, when the contraction norm is known (let's note it $|| \cdot ||_C$) there is no problem detecting the convergence since, by definition, we have:

$$\forall x, y, \quad ||G(x) - G(y)||_C \leq L||x - y||_C \quad \text{with} \quad L < 1 \tag{4.28}$$

which implies

$$||x^{(k+1)} - x^{(k)}||_C < ||x^{(k)} - x^{(k-1)}||_C \tag{4.29}$$

So, the distance between two global state vectors obtained from two consecutive iterations decreases according to the contraction norm $|| \cdot ||_C$. That distance between two consecutive iterations is often called the *residual* and, in some ways, represents the progression speed of the iterative process. Thus, when the right norm is used, the residual monotonously decreases toward zero, without reaching it if the convergence is asymptotic. However, when the residual becomes small enough, it can be assumed that the iterative process is sufficiently close to the exact solution to detect the convergence and stop the process. Hence, the residual is regularly compared to a given threshold defining a sufficiently small progression speed of the process to assume its stabilization. A schematic example of such a behavior is given in Figure 4.13.

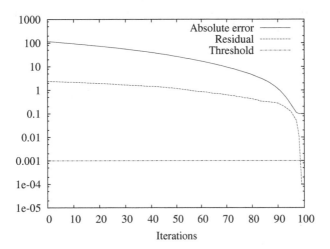

FIGURE 4.13: Monotonous residual decreases toward the stabilization according to the contraction norm.

Nevertheless, as mentioned above, it is not always possible to know the contraction norm and it is then necessary to arbitrarily choose one metric among all the possible ones. The most common metrics are the Euclidean norm:

$$||x^{(k)} - x^{(k-1)}||_2 = \sqrt{\sum_{i=1}^{n} (x_i^{(k)} - x_i^{(k-1)})^2} \tag{4.30}$$

and the max norm:

$$||x^{(k)} - x^{(k-1)}||_\infty = \max_i |x_i^{(k)} - x_i^{(k-1)}| \tag{4.31}$$

where $x_i^{(k)}$ and $x_i^{(k-1)}$ are the respective i^{th} component of state vectors $x^{(k)}$ and $x^{(k-1)}$.

So, when the norm used is not the contraction norm, nothing ensures that Equations (4.28) and (4.29) still hold. Using such an arbitrary norm makes the convergence detection far more difficult as there is no more valuable information about the position of the current state according to the exact solution. Thus, even when the residual becomes very small, nothing ensures us that the process is actually close to the exact solution. Typically, if the path followed by the iterative process toward the solution in the state space includes smaller variations than the chosen threshold according to the chosen norm, the convergence may be detected even though the current state may still be far from the solution. Moreover, even when the iterative process has a monotonous evolution toward the solution, i.e., when the distance between the current state vector and the exact solution always decreases from an iteration to the following one, the residual may not be monotonous. A schematic illustration of such a case is depicted in Figure 4.14.

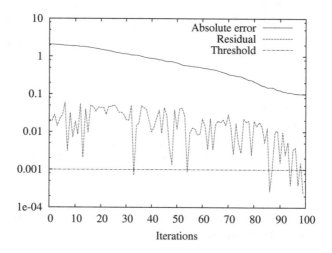

FIGURE 4.14: A monotonous error evolution and its corresponding non-monotonous residual evolution.

As can be seen, the problem induced by such variations of the residual is that important slow-downs like the one at iteration 33 may not correspond to the final stabilization of the process and lead to a false convergence detection if the threshold is set too high. On the other hand, if the threshold is set too low, the iterative process may not converge in reasonable time when the convergence is asymptotic. Some attempts have been made to overcome that problem by taking into account several consecutive residuals, for example:

$$residual^{(k)} = ||x^{(k)} - x^{(k-1)}|| + ||x^{(k-1)} - x^{(k-2)}|| \qquad (4.32)$$

in order to avoid false detections due to sharp slow-downs. However, that does not completely solve the problem since temporary slow-downs under the given threshold may have arbitrary lengths.

Thus, the choice of the norm used to compute the residual is a critical point in the design of iterative algorithms and the setting of its associated threshold often requires a careful analysis of the treated problem in order to ensure an appropriate convergence detection.

4.6 Exercises

1. Show that a nonsingular M-matrix has the form

$$sI - B,$$

   where $B \geq 0$ and $s > \rho(B)$.

2. Give examples of M-matrices and compute their principal minors.

3. Let A be a square matrix with $A_{i,j} \leq 0$ for $i \neq j$. Show that A is an M-matrix if and only if

$$A + \varepsilon I$$

   is a nonsingular M-matrix for all $\varepsilon > 0$.

4. Consider the two-point boundary-value problem:

$$-\frac{d^2u}{dx^2} = 4\pi^2 \sin 2\pi x, \ 0 \leq x \leq 1 \tag{4.33}$$

   and

$$u(0) = u(1) = 0.$$

   (a) Use the second central difference formulae with a constant step size $h = 1/(n+1)$ to approximate $\frac{d^2u}{dx^2}$ and show that the discrete approximation is the solution of a linear system $Au = b$ where $u = (u_1, ..., u_n)$.

   (b) For $n = 3$ show that

$$A = \begin{pmatrix} 2 & -1 & 0 \\ -1 & 2 & -1 \\ 0 & -1 & 2 \end{pmatrix}, \ b = \frac{\pi^2}{4} \begin{pmatrix} 1 \\ 0 \\ -1 \end{pmatrix}.$$

   (c) Use a direct method to solve the u_i.

   (d) Use the Jacobi and the Gauss-Seidel methods to find the discrete approximation.

(e) Compare the results to the true solution $u = \sin 2\pi x$ at $x = 1/4$, $x = 1/2$, $x = 3/4$ (Berman and Plemmons [31]).

5. Write a program to solve the two-point boundary-value problem (4.33) using Jacobi algorithm, SOR algorithm with optimum relaxation parameter for $n = 50$, $n = 200$, $n = 500$. Discuss the obtained results (Berman and Plemmons [31]).

6. Show that the system
$$\begin{cases} e^x - y = 0 \\ x - e^{-y} = 0 \end{cases}$$
has only one solution and write a program to solve it by the Newton method.

7. Write a program to solve the following system of nonlinear equations by the Newton method:
$$\begin{cases} x^2 + y^2 + z^2 - 3 & = 0 \\ xy + xz - 3yz + 1 = 0 \\ x^2 + y^2 - z^2 - 1 & = 0. \end{cases}$$

Study the convergence of the Newton method.

8. Consider the Laplace equation
$$\frac{\partial^2 f}{\partial x^2} + \frac{\partial^2 f}{\partial y^2} = 0, \quad (x,y) \in [0,2] \times [0,1],$$
with
$$f(0,y) = 0, \quad f(2,y) = 6$$
$$f(x,0) = 6, \quad \forall x \in [1,2]$$
$$\frac{\partial f}{\partial y}(x \le 1, y = 0) = 0, \quad \frac{\partial f}{\partial y}(x, y = 1) = 0.$$

(a) By using the finite difference method to approximate the second derivatives and following the illustration example of Chapter 1, write the linear system $Au = b$ whose solution coincides with the approximate solution of the above Laplace equation.

(b) By choosing step sizes $\Delta x = 2 \times 10^{-4}$ and $\Delta x = 10^{-4}$, write a program to solve the obtained linear system by the Jacobi algorithm and the Gauss-Seidel algorithm.

(c) Propose and write a program to solve the linear system by a multisplitting algorithm on 10 processors.

(d) Propose and write a program to solve the linear system by a two-stage multisplitting algorithm on 10 processors.

9. Consider the Poisson equation

$$-\Delta u = -32x(1-x)y(1-y) \; on \; \Omega =]0,1[^2 ,$$
$$u = 0 \; on \; \partial\Omega =]0,1[\times \{0,1\} \bigcup \{0,1\} \times]0,1[.$$

   (a) Following the above exercise, write a program to solve in parallel the discretized solution of this Poisson equation by the conjugate gradient method ($\Delta x = \Delta y = 10^{-4}$).

   (b) Propose and write a program to solve the linear system by different multisplitting algorithms on 10 processors.

   (c) Compare the overall times of the synchronous executions obtained with the different algorithms but with the same precision.

10. In all the algorithms presented in the implementation section, we have used the AllToAllV procedure that allows all processors to broadcast a part of a vector that they have computed. Write this procedure using only Send and Recv operations.

11. Implement all the algorithms presented in the implementation section using blocking and nonblocking receptions. Try to measure the performances with twenty processors or so.

12. Using an AllReduce operation allows us to simply diffuse the maximum of the local convergence on all processors. Try to implement the same thing using a master processor that will receive the local convergence of all processors, compute the global value and then diffuse the result to all processors.

13. With a sparse matrix, split into rectangular matrices as in Figure 4.6 on each processor, implement an algorithm that allows us to compute the arrays *DependsOnMe* and *IDependOn* as in Algorithm 4.6.

14. Most systems needing to be solved are sparse. Implement algorithms described in this chapter with a sparse matrix representation. Then compare the behavior of an algorithm optimized with a naive implementation.

Chapter 5

Asynchronous Iterations

Introduction

In the grid computing framework, especially when the clusters are distant, the ratio *computation time/communication time* can be weak and thus give a considerable importance to the communications. For this reason, powerful algorithms, such as those based on the minimization of a function, can paradoxically become less powerful in such environments. The synchronizations between the iterates provide the same convergence of algorithms as in the sequential case. Nevertheless, those synchronizations are penalizing in the distant clusters framework.

The asynchronous algorithms allow processors to compute at their own rhythms and to send the data when they become available. The communications as well as the iterations are desynchronized, avoiding the penalizing synchronizations and carrying out a kind of automatic overlapping of communications by computations. It is, however, necessary to precede any implementation of asynchronous iterative algorithms by a study of their convergence; this is due to the desynchronization of the iterations (notice that the study of the convergence is also necessary for all the iterative algorithms, even in the synchronous or the sequential framework).

In this chapter, we are interested in the multisplitting methods and their *two-stage* variants and in their coupling with the Newton method. Multisplitting algorithms allow us to carry out coarse grained parallelism which is very suited in the field of grid computing. Moreover, the two-stage multisplitting algorithms make it possible to choose, at the level of each processor, the best adapted sequential algorithm to the subproblem. We thus obtain coarse grained asynchronous algorithms with a coupling of different sequential algorithms.

5.1   Advantages of asynchronous algorithms

Contrary to synchronous implementations, in Asynchronous Iterations - Asynchronous Communication (AIAC) execution modes the processors are not coordinated in order to obtain a solution of a fixed point problem. Some processors are allowed to compute faster than others; some communications are allowed to be more frequent than others. The delays between processors are unpredictable and the transmission of messages may be accomplished in an unspecified order. Asynchronous iterations have been introduced in [38] by Chazan and Miranker for linear problems under the name *chaotic relaxation*. The pioneers in the study and the generalization to asynchronous algorithms are Miellou [86], Baudet [30], Robert, Charnay and Musy [101], Bertsekas and Tsitsiklis [32, 33], Bahi et al. [25] and Bahi [12].

The asynchronous iterations model describes a wide generalization of the successive approximation method in the case of a fixed point mapping defined on a product space, or even a product set.

This formulation is sufficiently general in order to contain:

- The successive approximation method which includes an inherent parallelism

- The Gauss-Seidel method which is often strictly sequential. The first of these two standard algorithms is well designed for parallelization while the second one is often, but not always, sequential. On the other hand, Gauss-Seidel iterations satisfy the so-called *Gauss principle*, which asserts that a new partial result is immediately used anew. Jacobi iterations do not satisfy the Gauss principle. The aim of asynchronous iterations is to satisfy in the best way the Gauss principle in a parallel framework. In some sense they afford a compromise between the usual good properties of the Jacobi and Gauss-Seidel methods.

Asynchronous executions have several potential advantages; we list some of them below (see [80]):

1. They reduce the effect of bottlenecks. Indeed, if for example the communication link between two processors is drastically slowed down then, contrary to synchronous executions, all the processors will go on and the two processors with the slow link will not slow down the processors which do not directly depend on them.

2. They reduce the synchronization penalty. A processor can compute the next iteration without waiting for the iterations computed by slower processors.

3. They are well designed for systems in which synchronizations are unrealistic such as very dense systems and for systems where global information is impossible to obtain such as decentralized systems.

4. They are easily restartable. For example, suppose that while solving an optimization problem, a change happens in one parameter, as may be the case in data networks. Then, while in synchronous computations the system has to be stopped and restarted, in asynchronous executions, the parameter is incorporated in each processor without waiting for all processors to do so.

5. They provide an improvement of the convergence thanks to the Gauss principle.

The major drawback of asynchronous algorithms is that they may diverge while their synchronous counterparts converge. Indeed, asynchronous iterations cannot be described mathematically by $x^{(k+1)} = T(x^{(k)})$.

Below, we give the mathematical model of asynchronous algorithms and we recall their convergence conditions.

5.2 Mathematical model and convergence results

5.2.1 The mathematical model of asynchronous algorithms

Suppose again that we have L processors and that an n-dimensional unknown vector is partitioned into L subvectors of dimension n_i, i.e., $n = \sum_{i=1}^{L} n_i$, so that each processor i can compute a vector of dimension n_i.

Consider the system of n equations

$$F(x) = 0, \qquad (5.1)$$

and suppose that (5.1) has a unique solution x^*. Suppose that after some algebraic transformations the above system of equations is rewritten as

$$x = T(x). \qquad (5.2)$$

Asynchronous executions of iterative algorithms associated to the above fixed point problem are described by the behavior of the following sequence (k denoting the kth iteration)

$$
\begin{cases}
Given\ x^{(0)} = (x_1^{(0)}, ..., x_L^{(0)}) \\
for\ k = 0, 1, 2... \\
\quad for\ i = 1, ..., L \\
\quad\quad x_i^{(k+1)} = \begin{cases} T_i(x_1^{(\rho_1^i(k))}, ..., x_i^{(\rho_i^i(k))}, ..., x_L^{(\rho_L^i(k))}) & if\ i \in s(k) \\ x_i^{(k)} & if\ i \notin s(k), \end{cases}
\end{cases}
\qquad (5.3)
$$

where $S = \{s(k)\}_{k \in \mathbb{N}}$ is a sequence of nonempty subsets of $\{1, ..., L\}$. The subsets $s(k)$ represent the set of components updated at the iteration k; it is usually called the *steering* of the algorithm.

For $i \in \{1, ..., L\}$, $\rho^i = \{\rho_1^i(k), ..., \rho_L^i(k)\}_{k \in \mathbb{N}}$ is a sequence of integers such that:

$$\forall i, j \in \{1, ..., L\}, \ \rho_j^i(k) \le k,$$

where $\rho_j^i(k)$ represents the iteration number of the data coming from processor j and available on processor i at iteration k. In other words, the quantity $k - \rho_j^i(k)$ represents the delay of processor j according to processor i when the latter processor computes the ith block at the kth iteration. That delay could be due to the communication time or to the computation time of the jth block.

The sequence (5.3) describes the behavior of iterative algorithms executed asynchronously on a parallel computer with L processors: at each iteration k, either the processor i computes $x_i^{(k+1)}$ by using the ith component T_i (so the ith component is updated) or it does not perform any computation and it keeps the value of $x_i^{(k)}$ computed at the previous iteration.

The two following conditions have to be ensured

$$card\{k \in \mathbb{N}, i \in s(k)\} = +\infty \tag{5.4}$$

and

$$\forall i, j \in \{1, ..., L\}, \ \lim_{k \to +\infty} \rho_j^i(k) = +\infty. \tag{5.5}$$

Condition (5.4) corresponds, in a standard way, to the fact that none of the equations is definitively forgotten and so, none of the corresponding components is not refreshed after a certain rank of the iterations.

Condition (5.5) implies that old information is purged from the system. This results from the fact that any information is transmitted during a delay shorter than the one of a finite number r of refreshment of components. So

$$\rho_j^i(k) = k - r$$

and (5.5) is satisfied.

It should be noted that in the model (5.3), we do not need to have knowledge of the delays nor of the updated components; we only have to be sure that conditions (5.4) and (5.5) are satisfied.

Two asynchronous executions of the same algorithm do not give rise to the same iterations, but when the convergence of an asynchronous algorithm is proved, this means that it converges whatever the actual conditions of experimentation, provided that conditions (5.4) and (5.5) are satisfied. In what follows, we will refer to asynchronous algorithms associated to the successive approximations generated by a fixed point mapping T by $(T, Async)$.

5.2.2 Some derived basic algorithms

Asynchronous fixed point methods are not only a family of algorithms suitable for asynchronous computations on multiprocessors, but also a general framework allowing a general formulation of iteration methods associated to a fixed point mapping on a product space, including the most standard ones such as the successive approximation method (linear or nonlinear Jacobi method) and linear or nonlinear Gauss-Seidel method among many others.

If we take $\rho_j^i(k) = k$ for all processors i and j and all $k \in \mathbb{N}$, then (5.3) describes synchronous parallel algorithms.

If we take $\rho_j^i(k) = k$ for all processors i and j, all $k \in \mathbb{N}$ and

$$s(k) = \{1, ..., L\} \ for \ k \in \mathbb{N}$$

then (5.3) describes the successive approximation method applied to (5.2).

If we take $\rho_j^i(k) = k$ for all processors i and j, all $k \in \mathbb{N}$ and

$$s(k) = \{1 + k(mod)L\} \ for \ k \in \mathbb{N}$$

then (5.3) describes the synchronous Gauss-Seidel algorithm.

Suppose that each of the L processors deals with m equations so that

$$Lm = n$$

then to modelize the Gauss-Seidel method executed synchronously on the L processors, it is sufficient to take

$$\begin{cases} \forall k \in \mathbb{N}, \ s(k) = \bigcup_{l=1}^{L} \{(l-1)\, m + 1 + k(mod)m\} \\ \forall i \in \{1, 2, ..., n\}, \rho_j^i(k) = k. \end{cases}$$

This method corresponds to a situation associated to Gauss's principle, which asserts that any obtained partial result is immediately used anew. This algorithm corresponds to an ideal case in which not only each processor synchronously swaps with the others the m equations to which it is devoted, but also synchronously accesses produced results, independently of the fact that they are produced by itself or by another.

Suppose now that each processor runs a sequential Gauss-Seidel algorithm along the m equations to which it is devoted, and that the L processors run asynchronously. If we suppose that each processor has a local steering

$$s^k(ql) = (l-1)\, n + 1 + ql(mod)m$$

then we get the asynchronous Gauss-Seidel method.

It is worthwhile to note that in the general situation of asynchronous algorithms we do not know the global steering $s(k)$, nor the delays $\rho_j^i(k)$ of such algorithms and so we are unable to specify their exact formulations. But to study the behavior of an asynchronous algorithm it is not important to know the exact expression of the steerings and delays; we only have to be sure that this algorithm admits such a formulation of the form (5.3) and that conditions (5.4), (5.5) are satisfied.

5.2.3   Convergence results of asynchronous algorithms

In the case of a Banach product space (see the Appendix), the main convergence theorem in a contraction framework is El Tarazi's Theorem [50], [51], formulated as follows. Consider the following maximum weighted norm defined on a product Banach space $E = \prod_{i=1}^{L} E_i$ by

$$
\begin{array}{c}
for\ u = (u_1, ..., u_L) \in E = \prod_{i=1}^{L} E_i \\
\|u\|_{\gamma,\infty} = \max_{1 \leq i \leq L} \frac{|u_i|_i}{\gamma_i}
\end{array}
\tag{5.6}
$$

where for $i \in \{1, ..., L\}$, $| \cdot |_i$ is a norm defined on E_i and γ_i are positive real numbers.

THEOREM 5.1
Let T be a mapping from $D(T) \subset E$ in E and suppose that:
(a) $D(T) = \prod_{i=1}^{L} D_i(T)$
(b) $T(D(T)) \subset D(T)$
(c) $\exists u^* \in D(T)$, such that $u^* = T(u^*)$
(d) $\forall u \in D(T), \|T(u) - u^*\|_{\gamma,\infty} \leq \beta \|u - u^*\|_{\gamma,\infty}$ with $0 < \beta < 1$,
then each asynchronous algorithm $(T, Async)$ associated to T converges to the fixed point u^* of T, whatever the starting point $u^0 \in D(T)$.

In fact, this theorem gives a generalization of a theorem due to Miellou [86]: Miellou's theorem is placed in a contraction matrix framework (with respect to a vectorial norm) which is included in the maximum weighted norm framework (with respect to a scalar norm). The weights γ_i are obtained by the Perron-Frobenius theorem in the case of monotone operators.

Bertsekas and Tsitsiklis established a more general convergence result based on nested sets [33]. This theorem is true in the general case of a Cartesian metric space. We give below its formulation in the case of the n-dimensional real space.

THEOREM 5.2
Let $E = \prod_{i=1}^{L} E_i \subset \prod_{i=1}^{L} \mathbb{R}^{n_i}$. Suppose that for each $i \in \{1, ..., L\}$, there exists a sequence of nested sets $E_i^{(k)}$ of subsets of E_i such that for all $k \geq 0$,

1. $E_i^{(k+1)} \subset E_i^{(k)}$,

2. $T(E^{(k)}) \subset E^{(k+1)}$, where $E^{(k)} = \prod_{i=1}^{L} E_i^{(k)}$,

then under assumptions (5.4) and (5.5) every limit point of the sequence $\left\{x^{(k)}\right\}_{k \in \mathbb{N}}$ generated by algorithm (5.3) and starting with $x^{(0)} \in E^{(0)}$ is a solution of the fixed point problem (5.2).

The last condition can be enlarged to the existence of attractors for the considered asynchronous iterations in the case of perturbed fixed point mapping, for example by round-off errors [87].

Convergence studies of asynchronous algorithms have been obtained by Chazan and Miranker [38] for linear systems, Miellou [86], Baudet [30], El Tarazi [50], [51] for contracting operators on the one hand and Miellou [86] and Bertsekas [32], for monotone iterations on the other hand. Uresin and Dubois [111] also have studied the convergence of asynchronous algorithms.

The sufficient conditions of Theorem 5.2 are shown to be necessary in the case where $n_i = 1$ for each i and T is a linear mapping ([38]).

Theorem 5.1 is in fact a particular situation which satisfies the general convergence result of Theorem 5.2. Indeed, if we consider the sets

$$E_i^{(k)} = \left\{ u_i \in \mathbb{R}^{n_i}, \ |u_i - u_i^*|_i \le \beta^k \left| u_i^{(0)} - u_i^* \right|_i \right\},$$

with $x^{(0)} \in D(T) \subset E$, then one can easily verify that the subsets $E_i^{(k)}$ satisfy the conditions of Theorem 5.2.

In several practical situations, the exact solution of the fixed point equation $x_i^{(k+1)} = T_i(x_1^{(\rho_1^i(k))}, ..., x_i^{(k+1)}, ..., x_L^{(\rho_L^i(k))})$ in (5.3) cannot be obtained or its computation may be prohibitive, so that one has to approximate this solution by performing some iterations of a convergent process. In these situations, we are dealing with inner and outer iterations. The iterations are called *nonstationary* since T depends on the current outer iteration k $(T^{(k)})$. Theorem 5.2 is still valid by replacing T by $T^{(k)}$. The obtained model includes the so-called *two-stage algorithms*.

When the new computed values are sent to other processors as soon as they are computed and without waiting for the convergence, we are in the context of asynchronous iterations with flexible communications [65], [49]. The mathematical model for these algorithms can be obtained by introducing a second set of delays as follows:

$$\begin{cases} Given \ x^{(0)} = (x_1^{(0)}, ..., x_L^{(0)}) \\ for \ k = 0, 1, 2... \\ \quad for \ i = 1, ..., L \\ \qquad x_i^{(k+1)} = \begin{cases} T_i \left((x_1^{(\rho_1^i(k))}, ..., x_L^{(\rho_L^i(k))}), (x_1^{(r_1^i(k))}, ..., x_L^{(r_L^i(k))}) \right) & if \ i \in s(k) \\ x_i^{(k)} & if \ i \notin s(k). \end{cases} \end{cases}$$

$$(5.7)$$

For a generalization of this model when the domain consists in multiple copies of E and each component of each copy is subject to different delays, see [64]. The following theorems ([65], [64]) which can be considered as a generalization of Theorem 5.1 give sufficient convergent conditions of asynchronous algorithms with flexible communications.

THEOREM 5.3

Assume that there exists $x^* \in E$ such that $T(x^*, x^*) = x^*$ and assume that there exist $\gamma \in [0, 1[$ and a weighted maximum norm such that for all $x, y \in E$,

$$\|T(x, y) - x^*\|_{\gamma, \infty} \leq \gamma \max(\|x - x^*\|_{\gamma, \infty}, \|y - x^*\|_{\gamma, \infty}),$$

then the asynchronous iterations described by (5.7) converges to x^*.

Asynchronous algorithms which satisfy conditions (5.4) and (5.5) are called *totally asynchronous algorithms* in opposition to partial asynchronous ones. The difference between the two kinds of algorithms mainly lies in the assumptions made on the delays between the processors. Condition (5.5) is replaced by the following conditions.

There exists a positive integer B such that for every iteration k, we have

$$\forall i, j \in \{1, ..., L\}, \ k - B + 1 \leq \rho_j^i(k) \leq k \tag{5.8}$$

$$\forall i \in \{1, ..., L\}, \ \rho_i^i(k) = k \tag{5.9}$$

Note that condition (5.8) means that old information is purged from the network after at most B iterations. This condition is satisfied in practice. Condition (5.9) means that the own computed components of a processor are never outdated. This last condition is also generally satisfied.

5.3   Convergence situations

5.3.1   The linear framework

Consider again the linear system

$$Ax = b, \tag{5.10}$$

where A is a $n \times n$ square nonsingular matrix and let

$$A = M - N \tag{5.11}$$

be a splitting of A, i.e., M is a nonsingular matrix. Consider the iterative algorithm associated to the splitting (5.11) and defined by

$$\begin{cases} x^{(0)} \ given \\ x^{(k+1)} = M^{-1}Nx^{(k)} + M^{-1}b, \ for \ all \ k \in \mathbb{N} \end{cases} \tag{5.12}$$

Let $T = M^{-1}N$ and $|T|$ denotes the matrix whose entries are the absolute values of the entries of T. Then we have the following result due to Chazan and Miranker [38].

THEOREM 5.4

1. If $\rho(|T|) < 1$ $(|T| = (|T_{i,j}|)_{i,j})$ then the asynchronous iterations (5.12) converge to the solution of (5.10).

2. If $\rho(|T|) \geq 1$ there exists a set of delays and strategies and an initial guess $x^{(0)}$ such that the corresponding asynchronous iteration does not converge to the solution of (5.10).

The proof of this theorem derives from the Perron-Frobenius theorem which states that $\rho(|T|) < 1$ if and only if T is a contraction with respect to a weighted maximum norm of the form (5.6).

In the next part of this section we are interested in the case of M-matrices (see the Appendix). Let us suppose that $A = (A_{i,j})$ is an M-matrix and consider the by point Jacobi splitting of A,

$$A = D - B,$$

where $D = diag(..., A_{i,i}, ...)$. Remark that as A is an M-matrix, B is a non-negative matrix. Consider the by point Jacobi iterations associated to the fixed point mapping G defined by

$$\forall x \in \mathbb{R}^n, \ y = G(x) \Leftrightarrow y = D^{-1}Bx + D^{-1}b,$$

then we have the following result which is proved for example in [26].

PROPOSITION 5.1
G is contractive with respect to a weighted maximum norm of the form (5.6), where the vector γ is obtained from the Perron-Frobenius theory for nonnegative matrices and the constant of contraction of G is either $\rho(D^{-1}B)$ if $D^{-1}B$ is irreducible or $\rho(D^{-1}B) + \varepsilon$ (for any $\varepsilon > 0$) if $D^{-1}B$ is reducible.

The next result, due to Bahi and Miellou [26], can be considered as an extension of the classical Stein-Rosenberg theorem [108] in the case of asynchronous algorithms. It allows us to build asynchronous convergent algorithms and to compare their speed of convergence.

Let $A = M - N$ be a regular per block splitting of A and T be the fixed point mapping corresponding to this splitting, i.e.,

$$\forall x \in \mathbb{R}^n, \ y = T(x) \Leftrightarrow y = M^{-1}Nx + M^{-1}b.$$

Consider also T_ω the relaxed fixed point mapping defined by

$$\forall x \in \mathbb{R}^n, \ y = T_\omega(x) \Leftrightarrow y = (1 - \omega)x + T(x),$$

then

PROPOSITION 5.2

If $\omega \in \,]0, \frac{2}{1+\rho(M^{-1}N)}[$ then any asynchronous algorithm associated to T_ω converges to the solution of (5.10). Moreover,

$$\rho(M^{-1}N) < \rho(D^{-1}B) < 1.$$

It should be noticed that, as stated in [26], the above proposition is also true in the case where A is an H-matrix with positive diagonal elements.

5.3.2   The nonlinear framework

Suppose that in the system of equations (5.1), F is a nonlinear mapping and that there exists a unique solution x^* on $D(F) \subset \mathbb{R}^n$. Suppose also that we have an iterative algorithm whose convergence is described by a fixed point mapping T which satisfies $x^* = T(x^*)$, then in [50] we have the following result on the local convergence of asynchronous iterations associated to T. This theorem can be considered as a generalization of the Ostrowski theorem [95] on successive iterations.

THEOREM 5.5

Assume that T is Fréchet differentiable at x^* and that x^* lies on the interior of $D(F)$. If $\rho(|T'(x)|) < 1$ then there exists a neighborhood V_{x^*} of x^* such that any asynchronous iteration associated to T and started with $x^{(0)} \in V_{x^*}$ converges to x^*.

An important iterative algorithm to solve (5.1) is the Newton algorithm for which

$$T(x) = x - F'(x)^{-1}F(x).$$

As $T'(x^*) = 0$, the above theorem can be applied.

In practice, the computation of $F'(x)^{-1}$ is expensive so one may use the quasi Newton methods by replacing $F'(x)^{-1}$ by a more simple computation. The above theorem is then applied by considering the spectral radius of the new $|T'(x)|$.

5.4   Parallel asynchronous multisplitting algorithms

In this section we come back to parallel multisplitting algorithms and we analyze their convergence when they are executed asynchronously on a grid environment. Multisplitting algorithms are well suited for parallel computing because each splitting gives rise to a subproblem which can be solved by a processor. Parallel asynchronous multisplitting algorithms may have several advantages. Indeed, in general, when the ratio communication time/computation time is not negligible, asynchronous execution may reduce the total time of computation by cancelling the idle times due to synchronizations between the iterations.

Another important point of multisplitting algorithms is that a processor may use a direct solver which is adapted to its problem independently from the other processors, which induces an interesting coupling of different solvers to treat a large problem on a grid environment.

In the next part of this section we follow the paper of Bahi et al. [27] and we give a unified mathematical framework of asynchronous multisplitting algorithms and then we consider the linear and nonlinear contexts.

5.4.1 A general framework of asynchronous multisplitting methods

We consider problems of the form (5.1)

$$F(x) = 0, \ x \in D \subset \mathbb{R}^n$$

where F is a nonlinear operator defined on a closed set D where

$$D = \prod_{i=1}^{n} d_i \text{ and } d_i \subset \mathbb{R} \text{ are closed and convex.} \qquad (5.13)$$

Suppose that the solution of (5.1) is x^*. Suppose also the existence of L mappings $T^{(l)}$ on D such that

$$T^{(l)}(D) \subset D \qquad (5.14)$$

and

$$T^{(l)}(x^*) = x^*, \ \forall l \in \{1, ..., L\} \qquad (5.15)$$

Assume that for $l \in \{1, ..., L\}$, $T^{(l)}$ is contractive with respect to x^* and to a norm $| \cdot |_{\infty,\gamma}$,

$$\begin{cases} |x|_{\infty,\gamma} = \max_{1 \le i \le n} \dfrac{|x_i|}{\gamma_i} \\ \gamma_i > 0. \end{cases} \qquad (5.16)$$

Here, $|x_i|$ denotes the absolute value of x_i in $\mathbb{R}$, i.e.,

$$\left| T^{(l)}(x) - x^* \right|_{\infty,\gamma} \le \nu_l \, |x - x^*|_{\infty,\gamma} \qquad (5.17)$$

DEFINITION 5.1   *A formal multisplitting associated to (5.1) is a collection of fixed point problems*

$$x - T^{(l)}(x) = 0, \ l \in \{1, ..., L\}$$

where each $T^{(l)}$ satisfies the conditions (5.14), (5.15) and (5.17). Let us fix the following notations,

$$x^l \text{ and } x^k, \ l, k \in \{1, ..., L\}$$

are vectors of $\mathbb{R}^n$ the components of which are

$$x_i^l \text{ and } x_j^k, \ i, j \in \{1, ..., n\}.$$

Define the extended fixed point mapping

$$\begin{cases} \mathcal{T} : (\mathbb{R}^n)^L & \longrightarrow \quad (\mathbb{R}^n)^L \\ X = (x^1, ..., x^L) & \longmapsto Y = (y^1, ..., y^L) \end{cases}$$

such that for $l \in \{1, ..., L\}$

$$\begin{cases} y^l = T^{(l)}(z^l) \\ z^l = \sum\limits_{k=1}^{L} E_{lk}(X)x^k \end{cases} \tag{5.18}$$

where $E_{lk}(X)$ are weighting matrices satisfying

$$\begin{cases} E_{lk}(X) \text{ are diagonal matrices} \\ E_{lk}(X) \geq 0 \\ \sum\limits_{k=1}^{L} E_{lk}(X) = I_n \ (\text{identity matrix in } \mathbb{R}^n), \ \forall l \in \{1, ..., L\} \end{cases} \tag{5.19}$$

Since $T^{(l)}(D) \subset D$ we have

$$\mathcal{T}(U) \subset U \tag{5.20}$$

where $U = \prod\limits_{i=1}^{L} D$.

Then the successive approximations associated to the extended fixed point mapping $\mathcal{T}$ describe the behavior of any multisplitting algorithms associated to (5.1).

As mentioned in [27] and in the previous chapter, the dependence of the weighting matrices $E_{lk}(X)$ on l, k and the current element X allows us:

- to take $E_{lk}(X) = E_k$ in order to obtain O'Leary and White multisplitting algorithms, as seen in Chapter 4.

- to define $E_{lk}(X) = E_{lk}$ depending on the index l in order to give a presentation of either the Schwarz alternating method or the general Schwarz multisplitting methods, as seen in Chapter 4.

- to take $E_{lk}(X)$ depending on both the index l and on the element X of $(\mathbb{R}^n)^L$, the value of which must be the current iterate X^p in order to describe two-stage multisplitting methods.

Under the assumptions of (5.14), (5.17), (5.19) we have the following result on T:

PROPOSITION 5.3

Denote $X^* = (x^*, ..., x^*)$ where x^* is the solution of (5.1), then T is contractive with respect to X^* and to $|\,.\,|_{\infty,\gamma}$ which is defined by

$$|X|_{\infty,\gamma} = \max_{1\leq k \leq L} \max_{1 \leq i \leq n} \frac{\left|\left(x^k\right)_i\right|}{\gamma_i} \tag{5.21}$$

Its constant of contraction is

$$\nu = \max_{1\leq l \leq L} \nu_l \tag{5.22}$$

and X^* is the fixed point of T.

PROOF Take any $Y = T(X)$, by (5.18) we have

$$\left|y^l - x^*\right|_{\infty,\gamma} = \left|T^{(l)}\left(\sum_{k=1}^{L} E_{lk}(X)x^k\right) - x^*\right|_{\infty,\gamma}$$

we have

$$\frac{\left|\left(\sum_{k=1}^{L} E_{lk}(X)\left(x^k - x^*\right)\right)_i\right|}{\gamma_i} = \frac{\left|\sum_{k=1}^{L} \sum_{j=1}^{n} (E_{lk}(X))_{i,j}\left(x^k - x^*\right)_j\right|}{\gamma_i}$$

Since the weighting matrices $E_{lk}(X)$ are diagonals, we have

$$\left|\sum_{j=1}^{n} (E_{lk}(X))_{i,j}\left(x^k - x^*\right)_j\right| = \left|(E_{lk}(X))_{i,i}\left(x^k - x^*\right)_i\right|$$

condition (5.19) gives

$$\left|\sum_{k=1}^{L}(E_{lk}(X))_{i,i}\left(x^k - x^*\right)_i\right| \leq \underbrace{\sum_{k=1}^{L}(E_{lk})_{i,i}(X)}_{1} \max_{1\leq k \leq L}\left|\left(x^k - x^*\right)_i\right|$$

so

$$\max_{1\leq i \leq n} \frac{\left|\left(T^{(l)}\left(\sum_{k=1}^{L} E_{lk}(X)x^k\right) - x^*\right)_i\right|}{\gamma_i} \leq \nu_l \max_{1\leq i \leq n} \max_{1\leq k \leq L} \frac{\left|\left(x^k - x^*\right)_i\right|}{\gamma_i}$$

then

$$\max_{1\le l\le L}\max_{1\le i\le n}\frac{\left|(y^l-x^*)_i\right|}{\gamma_i}\le\max_{1\le l\le L}\left(\nu_l\max_{1\le i\le n}\frac{\left|(x^l-x^*)_i\right|}{\gamma_i}\right)$$

by (5.21) and (5.22) we have

$$|Y-X^*|_{\infty,\gamma}\le\nu\,|X-X^*|_{\infty,\gamma}$$

since $\sum_{k=1}^{L}E_{lk}(X)=I_n$ and $x^*=T^{(l)}(x^*)$ we have $T(X^*)=X^*$. $\Box$

The above result gives the important following general convergence result of asynchronous multisplitting algorithms described by the fixed point mapping T for solving (5.1).

COROLLARY 5.1

Under the assumptions of Proposition 5.3, any asynchronous algorithm $(T, Async)$, corresponding to T and starting with $X^0 \in U$, converges to the solution of (5.1).

PROOF   Condition (5.20) implies that T has a unique fixed point; Theorem 5.1 and Proposition 5.3 end the proof. $\Box$

5.4.2   Asynchronous multisplitting algorithms for linear problems

In Chapter 4, we have shown how to build convergent synchronous multisplitting algorithms by splitting the nonsingular square matrix A of the linear system. We will now give a convergence result on asynchronous multisplitting algorithms for the solution of linear systems.

Consider again, as in Section 5.3.1, the linear system (5.10)

$$Ax = b,$$

where A is a $n \times n$ square nonsingular matrix and consider L splittings of A which are supposed to be regular

$$A = M_l - N_l, \ l = 1, ..., L$$

Then we can build a multisplitting as in definition (5.1) by setting

$$T^{(l)}(x) = M_l^{-1}N_lx + M_l^{-1}b.$$

Thus the successive approximations associated to the extended fixed point mapping T defined in (5.18) describe the behavior of parallel multisplitting algorithms for the solution of (5.10).

Process (5.3), where T is replaced by $\mathcal{T}$, describes the behavior of asynchronous execution of parallel multisplitting algorithms. A simple application of Proposition 5.2 shows that, for example, if A is an M-matrix then T_l are contractive mappings and as a consequence of Proposition 5.3 and Corollary 5.1 we obtain the following convergence result.

PROPOSITION 5.4
If the matrix A of the linear system (5.10) is an M-matrix, then any asynchronous multisplitting algorithm, associated with regular splittings of A, converges to the solution of the linear system (5.10).

By a suitable choice of the weighting matrices E_{lk} we can, as seen for the parallel synchronous case in Chapter 4, define asynchronous versions of the O'Leary and White and Schwarz multisubdomain algorithms.

5.4.3 Asynchronous multisplitting algorithms for nonlinear problems

In this section we introduce a tool which allows us to build splittings for either linear or nonlinear fixed point equations.

For a more general tool to build splittings of a nonlinear problem, the interested reader should consult [27]. We consider problems in the form (5.1),

$$F(x) = 0$$

which can be rewritten in the fixed point equation form (5.2),

$$x = T(x), \ x \in D \subset \mathbb{R}^n,$$

where D satisfies (5.13) and T satisfies the contraction assumption

$$\begin{cases} |T(x) - T(y)|_{\infty,\gamma} \le \nu \, |x - y|_{\infty,\gamma} \\ 0 < \nu < 1 \end{cases}$$

and

$$T(D) \subset D \qquad (5.23)$$

Let I_l, $l \in \{1, ..., L\}$, be subsets of $\{1, ..., n\}$ and I_l^C their complementaries

$$I_l \cup I_l^C = \{1, ..., n\}, \forall l \in \{1, ..., L\} \qquad (5.24)$$

and define the vectors $\sigma_i^l(u, v)$ by

$$\begin{cases} \sigma_i^l(u, v) = (w_1, ..., w_n) \text{ such that} \\ w_j = u_j \ if \ (i, j \in I_l) \ \text{or} \ (i, j \in I_l^C) \\ w_j = v_j \text{ otherwise} \end{cases} \qquad (5.25)$$

DEFINITION 5.2   *The block $\left(I_l, I_l^C\right)$ splittings are defined by the following mappings F_l*

$$\forall x \in D, \forall i \in \{1, ..., n\}, T_i^{(l)}(x) = T_i\left(\sigma_i^l\left(T^{(l)}(x), x\right)\right) \qquad (5.26)$$

In such a case we usually take

$$(E_{lk})_{i,j} = 0 \text{ or } (E_k)_{i,j} = 0 \text{ for } j \in I_k^C \qquad (5.27)$$

It should be noticed that if we except particular problems which admit a natural block decomposition structure suitable for block iterative algorithms, the previous condition (5.27) is very important, especially for overlapping block decomposition techniques, because in the evaluation of $T^{(l)}$, for any k we never have to use any component the index of which lies in I_k^C, so in the block $\left(I_l, I_l^C\right)$ splitting the solution of a diagonal block subproblem associated to any I_l^C never has to be computed.

PROPOSITION 5.5
For $l \in \{1, ..., L\}$, $T^{(l)}$ is $|\,.\,|_{\infty,\gamma}$ contractive, its constant is less than or equal to the constant of T and the fixed point of $T^{(l)}$ is x^*.

PROOF

$$\frac{\left|T_i^{(l)}(x) - T_i^{(l)}(y)\right|}{\gamma_i} = \frac{\left|T_i\left(\sigma_i^l\left(T^{(l)}(x), x\right)\right) - T_i\left(\sigma_i^l\left(T^{(l)}(y), y\right)\right)\right|}{\gamma_i}$$

$$\leq \nu\left|\sigma_i^l\left(T^{(l)}(x), x\right) - \sigma_i^l\left(T^{(l)}(y), y\right)\right|_{\infty,\gamma}$$

$$\leq \nu \max\left(\max_{1 \leq j \leq n} \frac{\left|T_j^{(l)}(x) - T_j^{(l)}(y)\right|}{\gamma_j}, \max_{1 \leq j \leq n} \frac{|x_j - y_j|}{\gamma_j}\right)$$

so either

$$\frac{\left|T_i^{(l)}(x) - T_i^{(l)}(y)\right|}{\gamma_i} \leq \nu \max_{1 \leq j \leq n} \frac{\left|T_j^{(l)}(x) - T_j^{(l)}(y)\right|}{\gamma_j}$$

which implies that $\left|T_i^{(l)}(x) - T_i^{(l)}(y)\right| = 0$
or

$$\frac{\left|T_i^{(l)}(x) - T_i^{(l)}(y)\right|}{\gamma_i} \leq \nu \max_{1 \leq j \leq n} \frac{|x_j - y_j|}{\gamma_j}$$

so

$$\left|T^{(l)}(x) - T^{(l)}(y)\right|_{\infty,\gamma} \leq \nu\,|x - y|_{\infty,\gamma}$$

which implies that $T^{(l)}$ is contractive and that its constant is less than or equal to ν.

Moreover by (5.23) F and T have an unique fixed point; let

$$T_i^{(l)}(x) = x_i$$

so equivalently

$$x_i = T_i\left(\sigma_i^l(x, x)\right)$$

so

$$x = T(x)$$

hence $T^{(l)}$ and T have the same fixed point x^*. $\qquad\qquad$ ▯

5.4.3.1 Extended fixed point mapping associated with $\left(I_l, I_l^C\right)$ multisplitting

Take the diagonal positive matrices $E_{lk}(X)$ depending only on k

$$E_{lk}(X) = E_k$$

and satisfying

$$\begin{cases} \sum\limits_{k=1}^{L} E_k = I_n \\ (E_k)_{i,i} = 0, \ \forall i \notin I_k \end{cases} \qquad (5.28)$$

The asynchronous iterations corresponding to $\left(I_l, I_l^C\right)$ multisplitting are defined by the fixed point mapping

$$\mathcal{T}^{OW}(x^1, ..., x^L) = (y^1, ..., y^L) \ such \ that$$

$$\begin{cases} y^l = T^{(l)}(z) \\ z = \sum\limits_{k=1}^{L} E_k x^k \end{cases} \qquad (5.29)$$

where for $l \in \{1, ..., L\}$, $T^{(l)}$ is defined by (5.26). We remark that this multisplitting algorithm is analogous to O'Leary and White multisplitting algorithms for nonlinear problems.

As a consequence of Propositions 5.3 and 5.5 we have

COROLLARY 5.2

Any asynchronous algorithm $(\mathcal{T}^{OW}, Async)$ corresponding to $\mathcal{T}^{OW}$ and starting with $X^0 \in U$ converges to the solution of (5.1).

5.4.3.2 The discrete analogue of Schwarz alternating method and its multisubdomain generalizations

Asynchronous Schwarz alternating methods and their multisubdomain generalizations are obtained by choosing the weighted matrices exactly as in Chapter 4. In the following, we point out, once again, these choices.

5.4.3.3 Discrete analogue of the Schwarz alternating method

Suppose $I_1 \cap I_2 \neq \emptyset$, so we have an overlap between the 1st and the 2nd subdomains. Consider the matrices E_{lk} such that

$$(E_{11})_{i,i} = \begin{cases} 1 \ \forall i \in I_1 \\ 0 \ \forall i \notin I_1 \end{cases}, \quad (E_{12})_{i,i} = \begin{cases} 0 \ \forall i \in I_1 \\ 1 \ \forall i \notin I_1 \end{cases} \tag{5.30}$$

$$(E_{21})_{i,i} = \begin{cases} 1 \ \forall i \notin I_2 \\ 0 \ \forall i \in I_2 \end{cases}, \quad (E_{22})_{i,i} = \begin{cases} 0 \ \forall i \notin I_2 \\ 1 \ \forall i \in I_2 \end{cases}$$

Define the fixed point mapping

$$\mathcal{T}^S(x^1, x^2) = (y^1, y^2) \ such \ that \ for \ l = 1, 2$$

$$\begin{cases} y^l = T^{(l)}(z^l) \\ z^l = \sum_{k=1}^{2} E_{lk} x^k \end{cases} \tag{5.31}$$

where for $l \in \{1, 2\}$, $T^{(l)}$ is defined by (5.26). Then the additive discrete analogue of the Schwarz alternating method corresponds to the successive approximation method applied to $\mathcal{T}^S$, and the multiplicative discrete analogue of the Schwarz alternating method corresponds to the block nonlinear Gauss-Seidel method applied to $\mathcal{T}^S$. For the use of such methods as preconditioners of Krylov spaces methods, we refer to [72], [106].

5.4.3.4 Discrete analogue of the multisubdomain Schwarz method

We introduce the weighting matrices E_k satisfying (5.28) and the matrices E_{lk} such that for $l \in \{1, ..., L\}$

$$(E_{ll})_{i,i} = \begin{cases} 1 \ if \ i \in I_l \\ 0 \ if \ i \notin I_l \end{cases}$$
$$(E_{lk})_{i,i} = \begin{cases} 0 \ if \ i \in I_l \\ (E_k)_{i,i} \ if \ i \notin I_l \end{cases} \tag{5.32}$$

the asynchronous iterations, corresponding to the discrete analogue of the multisubdomain Schwarz method, are defined by the fixed point mapping $\mathcal{T}^{MS}$

$$\mathcal{T}^{MS}(x^1, ..., x^L) = (y^1, ..., y^L) \ such \ that$$

$$\begin{cases} y^l = T^{(l)}(z^l) \\ z^l = \sum_{k=1}^{L} E_{lk} x^k \end{cases} \tag{5.33}$$

where E_{lk} are defined by (5.32) and $T^{(l)}$ are defined by (5.26).

$\mathcal{T}$ being $\mathcal{T}^{OW}$ or $\mathcal{T}^S$ or $\mathcal{T}^{MS}$ we have the following Corollary.

COROLLARY 5.3

Any asynchronous algorithm $(\mathcal{T}, Async)$, corresponding to $\mathcal{T}$ and starting with $X^0 \in U$, converges to the solution of (5.1).

5.5 Coupling Newton and multisplitting algorithms

The standard algorithm for solving the system of nonlinear equations (5.1) is the Newton algorithm; an effective way to use the Newton algorithm in a parallel environment is to couple it with multisplitting algorithms.

There are two ways to realize this coupling. The first one consists in splitting the linear problems involved in each iteration of the Newton algorithm and the second one consists in splitting the nonlinear problem (5.1) itself into subproblems and solving each subproblem using the Newton algorithm. Below, we describe the algorithmic formulation of these two kinds of mixed Newton multisplitting algorithms.

5.5.1 Newton-multisplitting algorithms: multisplitting algorithms as inner algorithms in the Newton method

Recall that the Newton algorithm for solving the nonlinear system of equation (5.1), $F(x) = 0$ is described by the iterations

$$x^{(k+1)} = x^{(k)} - F'(x^{(k)})^{-1}F(x^{(k)}), \ k = 0, 1, 2, ...$$

As in Chapter 4, we will suppose that (5.1) has a solution x^*, that F is Fréchet differentiable on a neighborhood of x^* and that F' is nonsingular and Lipschitz continuous on a neighborhood of x^*. We have seen in Chapter 4 that the Newton method involves the solution of a linear system

$$F'(x^{(k)})y = F(x^{(k)}) \tag{5.34}$$

and that this solution allows the computation of the next Newton iterates $x^{(k+1)}$ by setting $y^{(k)} = y$ in the following equation:

$$x^{(k+1)} = x^{(k)} - y^{(k)}, \ k = 0, 1, 2, ...$$

The solution of (5.34) by splitting $F'(x^{(k)})$ gives rise to the multisplitting methods to solve this kind of problem. We call the global algorithm to solve (5.1) the *Newton-multisplitting* algorithm.

So, suppose we have L processors and that, as explained in Chapter 4, we have L splittings of $F'(x^{(k)})$ at each iteration k, so that we have

$$F'(x^{(k)}) = M_l(x^{(k)}) - N_l(x^{(k)}), \ l = 1, ..., L. \tag{5.35}$$

For simplicity sake, suppose that the weighting matrices only depend on one index and that the solution of system (5.34) is approximated by performing q iterations of the multisplitting method.

The parallel Newton-multisplitting method can be defined as follows

$$x^{(k+1)} = G(x^{(k)}), \tag{5.36}$$

where

$$G(x) = x - A(x)F(x), \tag{5.37}$$

and

$$A(x) = \sum_{l=1}^{L} E_l(x) \sum_{j=0}^{q-1} (M_l(x)^{-1} N_l(x))^j M_l(x)^{-1}.$$

If we take $y^{(0)} = 0$, then

$$A(x) = \sum_{l=1}^{L} E_l(x)(I - (M_l(x)^{-1} N_l(x))^q (F'(x))^{-1}. \tag{5.38}$$

THEOREM 5.6

If the splittings (5.35) are weak regular convergent, then there exists a neighborhood V_{x^*} of the solution x^* such that any asynchronous Newton-Multisplitting algorithm associated to (5.37) and (5.38), and starting from $x^{(0)} \in V_{x^*}$ converges to x^*.

PROOF   We apply Theorem 5.5. We have

$$G'(x^*) = I - A(x^*)F'(x^*). \tag{5.39}$$

From (5.38) we have

$$G'(x^*) = I - \sum_{l=1}^{L} E_l(x^*)(I - (M_l(x^*)^{-1} N_l(x^*))^q. \tag{5.40}$$

The properties of the weighting matrices imply that

$$|G'(x^*)| = G'(x^*) = \sum_{l=1}^{L} E_l(x^*)(M_l(x^*)^{-1} N_l(x^*))^q. \tag{5.41}$$

As the splittings (5.35) are convergent, we deduce by the application of Proposition 3.2 of [27] that

$$\rho(G'(x^*)) \leq \max_{1 \leq l \leq L} \rho((M_l(x^*)^{-1} N_l(x^*))^q) < 1.$$

The result follows from Theorem 5.5.   □

There exist particular situations which satisfy the assumptions of the above convergence result. For example, if $F'(x)$ is monotone (i.e., $F'(x)^{-1} \geq 0$) then every weak regular splitting of $F'(x)$ is convergent [31].

5.5.2 Nonlinear multisplitting-Newton algorithms

Another way to mix the Newton method and the multisplitting approach is to use the result of Bahi et al. [26], [27] on nonlinear multisplitting. Indeed, the Newton method applied to the nonlinear problem (5.1) is described by the iterations associated to the contractive fixed point mapping

$$x = T(x), \ x \in \mathbb{R}^n,$$

where

$$T(x) = x - F'(x)^{-1} F(x).$$

Consider now L subsets I_l of $\{1, ..., n\}$ and weights E_{lk}, $l, k \in \{1, ..., L\}$, then Definition 5.2 of Section 5.4.3 allows us to generate L splittings of (5.2),

$$x = T^{(l)}(x), \ x \in \mathbb{R}^n,$$

with

$$x_i = T_i \left(\sigma_i^l \left(T^{(l)}(x), x \right) \right).$$

The application of Proposition 5.5 implies that $T^{(l)}$ are contractive mappings, so they define a nonlinear formal multisplitting: $x - T^{(l)}(x), l \in \{1, ..., L\}$ as in Definition 5.1. We consider asynchronous algorithms associated to those splittings and weights E_{lk}. We call those algorithms *nonlinear multisplitting-Newton* algorithms.

Practically, the iterations, generated by each fixed point mapping $T^{(l)}$ defined just above, correspond to the iterations generated by the Newton algorithm and applied to a subproblem of (5.1). These subproblems are defined by the (I_l, I_l^C) splittings; they correspond to the computation of $card(I_l)$ components of x. We then have the convergence result which is a consequence of Proposition 5.3 and Corollary 5.1.

PROPOSITION 5.6

Suppose that the Newton algorithm with the initial guess $x^{(0)} \in V_{x^*}$ converges to x^*, a solution of (5.1) in $V_{x^*} \subset D(F)$, then the nonlinear multisplitting-Newton algorithm started with $x^{(0)}$ converges to x^*.

5.6 Implementation

The implementation of an asynchronous iterative algorithm may seem easier to achieve since there is no more synchronization. Nevertheless, as described in the previous section, the convergence detection is different and is not the most trivial point to implement in a distributed environment. Indeed, according to the dedicated architecture, a centralized mechanism can either be used (for a parallel architecture or a cluster with a high speed network with a quite limited number of processors) or is completely inconceivable (with a distributed cluster or a grid with a large number of processors). In Section 3.2.1, the classification of parallel iterative algorithms points out another crucial point allowing us to distinguish a synchronous parallel iterative algorithm from an asynchronous one. It deals with the communications management. With a synchronous algorithm, all messages sent are received and used. With an asynchronous algorithm, according to the implementation of communications management, it may not be the case. Indeed, on the sending side, some messages may not be actually sent if newer local data are available before their emission. Likewise, on the receiving side, some messages may not be taken into account if newer messages arrived before their use. Furthermore, it should be remembered that asynchronous iterative algorithms support message loss.

In Chapter 4, we have detailed the parallelization of some well-known algorithms. Only some of them can be executed using asynchronous iterations. For example, it is not possible to execute a parallel Conjugate Gradient, or a GMRES with asynchronous iterations. Roughly speaking, only algorithms based on the Jacobi method and the multisplitting method can be executed in an asynchronous mode.

Before explaining those algorithms, it is essential to review the different ways to manage the asynchronism in programming and execution environments. With AIAC algorithms, the iterations are asynchronous and so are the communications. Consequently, communications must be dissociated from the computations. With several traditional parallel environments based on the message passing paradigm (like PVM [66], MPI [71]), it is possible to use buffered sendings and nonblocking receptions. Nevertheless all emitted messages must be received using a receive operation. One of the particularities of AIAC algorithms is that when there are several versions of the same message (corresponding to different iterations), the program should only take the last version in order to converge faster. To clarify this, let us take a simple example. Consider that two processors are executing an AIAC algorithm and that processor 1 performs its iterations two times faster than processor 2. Consider also, that at each iteration processor 2 receives on average two messages from processor 1 and that processor 1 only receives a message from processor 2 approximately one iteration out of two. If the processors only test an iteration once, if a message arrives, then processor 2 would have a lot of delay in the reception of messages since at each iteration k it would approximately have k messages in delay. Of course, it is not possible to know a priori the number of messages that a processor will receive at each iteration and this number varies from one iteration to another. So, in a traditional message passing based environment, a naive solution consists in receiving all messages at each iteration and only using the last one. Then, the problem comes from the convergence detection that must be very efficient, and for that, as soon as a message for the convergence arrives it must be detected. That is why it is essential to dissociate the communications from the computations. For that, the only solution from our point of view lies in using a multithreaded environment which allows us to execute the computations in one thread and the management of the communications in other threads.

In order to keep the same formalism, we do not focus on the implementation of AIAC algorithms with shared memory architectures. With such environments, as soon as a mechanism to simulate communications between AIAC algorithms has been implemented, the following algorithms are quite easy to adapt.

5.6.1 Some solutions to manage the communications using threads

According to the flexibility of the communications in an AIAC algorithm, it is possible to distinguish different levels of communications management. The simplest solution, from the programmer point of view, consists in using an environment suited to the design of AIAC algorithms. Currently two programming environments fulfill those requirements, namely, JACE [23, 22, 24, 19] and CRAC [40]. Both environments have been developed in order to provide a communication library that allows us to design synchronous and asynchronous iterative algorithms. They use two queues that are executed into two threads: one for the message sendings and another one for the receptions. According to the execution mode (synchronous or asynchronous), the operating of those queues is different. In the synchronous mode, those queues are managed traditionally, i.e., when a computation task needs to send a message, the message is put in the sending queue that actually sends it, the reception queue receives it on the destination processor and the computation task on the other machine can use the message. In the asynchronous mode, the sending queue first checks whether a similar version of the message is not already in the queue (based on its tag, sender and receiver). In this case, the previous one is replaced by the newest one. The reception queue acts similarly when receiving a message. It checks the reception queue and replaces an old message by a recent one whenever possible. So, when a computation task receives a message, it is ensured to have the latest version available. Of course, in JACE and CRAC, the programmer does not need to interact with the threads which transparently manage the communications. Those two environments are detailed in Sections 6.2.1 and 6.2.2.

With a multithreaded version of MPI [7], or Corba [98] or PM2 [89] which are implicitly multithreaded, it is possible to implement AIAC algorithms. However, this requires a stronger endeavor from the programmer point of view since the management of threads is explicit. Using an environment with explicit management of threads, a programmer may assign one or more threads in charge of sending some messages and as many in charge of receiving them. Although the endeavor is stronger, it allows us to manage more precisely the communications. For example, it is possible to implement flexible models defined in Section 3.4.3. The sending of a message is as flexible as using any message passing interface since a user can send a message anywhere in a program. However, using a thread that can directly handle a message as its reception occurs is a possible source of convergence speed-up. As network resources in a distant environment often are a critical point, a possible strategy consists in assigning a thread to each destination neighbor and in waiting for the previous message to have arrived at its destination before sending another one. For that, the use of a mutex combined with an acknowledgment message allows the programmer to control the sending of each message. The principle is the following: when a processor wants to send a message to a given neighbor, the thread that is dedicated to this neighbor is locked (unless it was already locked and in that case, the message is not sent). Then the message is sent, the thread on the emitter processor is blocked until the neighbor confirms that it has received the message. When the emitting thread has received the acknowledgement of reception, it unlocks the mutex. So, it is then ready to send another message. If another message was supposed to be sent, then the mutex would be locked so the sending would not be possible. As a consequence, the network would not be overloaded with a useless message, since the previous one would not have been handled yet. One of the drawbacks of this explicit management of threads is that when the number of neighbors per processor is not known in advance or is dynamic, it is difficult to define a number of threads a priori. Moreover, that difficulty comes for both the sending and the reception. Furthermore, the explicit management of threads requires much more attention than traditional programming because they may lead to deadlock situations if the programmer is not very attentive. When the number of threads running simultaneously becomes too important, the scheduling may be less fair, which is not acceptable. In fact, the fairness is an essential requirement of the threads management since the convergence conditions of AIAC algorithms involve that each processor should be able to regularly update its components.

In the following we present some asynchronous iterative algorithms in which we consider that messages arrive in their emitted order. If this is not the case, a simple mechanism should be added which consists in adding the iteration number at which the sent data have been produced on the sender. Then, on the receiver, the iteration number included in the message is compared to the one of the last message taken into account from that source. Finally, if the number in the message is smaller than the current one, the message is suppressed without being taken into account. Otherwise, the message is used and the current iteration number related to that source is updated. Implementing that mechanism allows us to ensure a faster convergence.

5.6.2 Asynchronous Jacobi algorithm

The asynchronous version of the synchronous Jacobi algorithm presents several similarities with it. In fact only two parts are different, the management of the communications and the convergence detection. In Algorithm 5.1 we give a possible implementation of the asynchronous Jacobi. With such a formalism which hides the mechanism to manage the asynchronism, it is quite easy to write the asynchronous version starting with the synchronous one. In this algorithm, receptions are nonblocking whereas they were blocking in the synchronous version. So, after the sendings, a processor takes the last version of its neighbors' messages if new messages have arrived since the previous iteration. In Algorithm 5.1, restarting a new iteration without receiving any new messages leads to the same computation. That is why a simple way to enhance this algorithm consists in detecting if a new message has arrived at each iteration. If this is not the case, it would probably be better to wait for a few micro seconds and to test again if a new message has arrived. According to the number of neighbors, it may be wise to wait for the reception of a given number of messages.

The other difference with the synchronous version of the Jacobi algorithm concerns the convergence detection. As described in Section 5.7, it is possible to use a centralized version or a decentralized version in order to detect the convergence. In this algorithm we consider that a function called *convergence* allows us to detect the global convergence. This function uses the local error and the threshold *Epsilon* used to stop the iterations.

If receptions are directly managed by threads, they can occur at any moment in the program. In that case, Algorithm 5.1 is slightly different. In fact, the call to the receive function is no longer in the main iteration. As receptions are completely free and do not only occur at the end of an iteration, this version of the Jacobi algorithm is based on what we have called *receiver-side semi-flexibility* (cf Section 3.4.3.1.2).

5.6.3 Asynchronous block Jacobi algorithm

This algorithm is the asynchronous version of the synchronous block Jacobi one. Compared to the synchronous version, it presents the advantage of being less perturbed by synchronizations as each processor has a block of the matrix to compute with a direct method that may require a non-negligible time for each subsystem. Compared to the Jacobi algorithm (without blocks), the asynchronous block version will probably converge in less iterations but they will probably be longer. So, during the solving of the subsystem, messages from the neighbors have time to arrive. Consequently, the overlapping of messages by computations with this algorithm may be more important than in the asynchronous Jacobi algorithm, especially when dealing with large matrices.

This algorithm, like its synchronous version, also requires that the linear solver used to solve the subsystem provides a result without approximation. In other words, a direct method is required.

Algorithm 5.1 Asynchronous Jacobi algorithm

NbProcs : number of processors
MyRank : rank of the processor
Size : local size of the matrix
SizeGlo : global size of the matrix
Offset : offset of the global index
A[Size][SizeGlo]: local part of the matrix
X[Size]: local part of the solution vector
XOld[SizeGlo]: global solution vector
B[Size]: local part of the right-hand side vector
Error : local error
Epsilon: desired accuracy
Converged: convergence state

repeat
  for i=0 to Size−1 **do**
    X[i] ← 0
    for j=0 to i+Offset−1 **do**
      X[i] ← X[i]+A[i][j]×XOld[j]
    end for
    for j=i+Offset+1 to SizeGlo−1 **do**
      X[i] ← X[i]+A[i][j]×XOld[j]
    end for
  end for
  for i=0 to Size−1 **do**
    X[i] ← (B[i]−X[i])/A[i][i+Offset]
  end for
  Error← 0
  for i=0 to Size−1 **do**
    Error ← max(Error, abs(A[i]−XOld[i+Offset]))
    XOld[i+Offset] ← X[i]
  end for
  for k=0 to NbProcs−1 **do**
    if k ≠ MyRank **then**
      Send(k, X)
    end if
  end for
  for k=0 to NbProcs−1 **do**
    if k ≠ MyRank **then**
      Recv(k, XOld[k×Size])
    end if
  end for
  Converged ← convergence(Error, Epsilon)
until Converged = true

Algorithm 5.2 Asynchronous block Jacobi algorithm

NbProcs: number of processors
MyRank: rank of the processor
Size: local size of the matrix
SizeGlo: global size of the matrix
Offset: offset of the global index
A[Size][SizeGlo]: local part of the matrix
X[Size]: local part of the solution vector
B[Size]: local part of the right-hand side vector
BTmp[Size]: intermediate local part of the right-hand side vector
XOld[SizeGlo]: global solution vector
Error: local error
Epsilon: desired accuracy
Converged: convergence state

repeat
    for i=0 to Size−1 **do**
      BTmp[i]← B[i]
    end for
    for i=0 to Size−1 **do**
      for j=0 to Offset−1 **do**
        BTmp[i] ← BTmp[i]−A[i][j]×XOld[j]
      end for
      for j=Offset+Size to SizeGlo−1 **do**
        BTmp[i] ← BTmp[i]−A[i][j]×XOld[j]
      end for
    end for
    X← Solve(A, BTmp)
    Error← 0
    for i=0 to Size−1 **do**
      Error ← max(Error, abs(X[i]−XOld[i+Offset]))
      XOld[i+Offset]← X[i]
    end for
    for k=0 to NbProcs−1 **do**
      if k ≠ MyRank **then**
        Send(k, X)
      end if
    end for
    for k=0 to NbProcs−1 **do**
      if k ≠ MyRank **then**
        Recv(k, XOld[k×Size])
      end if
    end for
    Converged ← convergence(Error, Epsilon)
until Converged = true

5.6.4   Asynchronous multisplitting algorithm for solving linear systems

The asynchronous version of the multisplitting method for solving linear systems is designed to be efficient for grid or distant clusters. This method actually features interesting characteristics for this. It is a coarse grained algorithm since a processor solves the subsystem it is in charge of at each iteration either using a sequential iterative solver (i.e., so we obtain a two-stage algorithm) or a direct one. According to the characteristics of the subsystems obtained by the splitting and the parameters of the architecture, a good choice of the inner method can drastically change the performances. This method allows us to overlap communications with computations. This feature is typically provided by the asynchronism of the method. Consequently, we strongly believe that this method is particularly well suited to solve large linear systems in grid environments. Compared to the synchronous version, the asynchronous one only has two modifications. Those two modifications concern the two main differences between a synchronous and an asynchronous version of the same algorithm for which the convergence proof in the asynchronous mode has been previously studied, that is to say, the management of the communications and the convergence detection, as previously mentioned in this section.

With this method, it is strongly recommended to count the number of messages received per iteration and to take into account this number in order to decide if the program should wait for other messages or run the next iteration. As previously mentioned, running a new iteration without any new message will produce the same result, which is not interesting from the computational point of view. In order to increase the convergence speed it is sometimes more interesting to wait for a small span of time, for example 1 ms, to receive some new messages rather than using only one new message before running the next iteration. In the synchronous version of this algorithm we have presented the multiple ways of overlapping some components. In Algorithm 5.4 we present the small changes in the multisplitting algorithm in order to take into account the *Overlap* components which are overlapped. Obviously, we consider that the size of the *Overlap* parameter is less than the size of the subsystem.

Using the overlapping of components has two main impacts on the execution of an AIAC algorithm. The first one is that the number of iterations required to reach the convergence threshold is smaller. That is the positive point. The second impact, which is a drawback, is that the size of each subsystem is larger, and consequently, the time to solve a subsystem is longer. That is why using the overlapping mechanism may reduce the number of iterations when this number is high, i.e., the spectral radius of iteration matrix is close to one. Nevertheless, according to the method used to solve subsystems, the solving time may change. If a direct method is used, then one of the most time-consuming tasks consists in factorizing the matrix. At each iteration of the multisplitting method, only the right-hand side changes, so the factorized

---

Algorithm 5.3 Asynchronous linear multisplitting algorithm

NbProcs: number of processors
MyRank: rank of the processor
Size: local size of the matrix
SizeGlo: global size of the matrix
Offset: offset of the global index
A[Size][Size]: local block-diagonal part of the matrix
DepLeft[Size][Offset]: submatrix with left dependencies
DepRight[Size][SizeGlo-Offset-Size]: submatrix with right dependencies
DependsOnMe[NbProcs]: array of the dependent processors
IDependOn[NbProcs]: array of the processors this processor depends on
B[Size]: right-hand side vector of the subsystem
X[Size], XOld[Size]: local part of solution vectors of the subsystem
XLeft[Offset]: left part of the solution vector of the system
XRight[SizeGlo-Offset-Size]: right part of the solution vector of the system
BLoc[Size]: array containing the local computations on the right-hand side
TLoc[Size]: array used for the receptions of the dependencies
Error: local error
Epsilon: desired accuracy
Converged: convergence state

repeat
    BLoc ← B
    if MyRank≠0 **then**
        BLoc ← BLoc−DepLeft×XLeft
    end if
    if MyRank ≠ NbProcs−1 **then**
        BLoc ← BLoc−DepRight×XRight
    end if
    X ← Solve(A, BLoc)
    for i=0 to NbProcs−1 **do**
        if i ≠ MyRank and DependsOnMe[i] **then**
            Send(i, PartOf(X, i))
        end if
    end for
    for i=0 to NbProcs−1 **do**
        if i ≠ MyRank and IDependOn[i] **then**
            if Recv(i, TLoc) **then**
                Update XLeft or Xright with TLoc according to the processor *i*
            end if
        end if
    end for
    Error← 0
    for i=0 to Size−1 **do**
        Error ← max(Error, abs(X[i]−XOld[i]))
        XOld[i]← X[i]
    end for
    Converged ← convergence(Error, Epsilon)
until Converged = true

Algorithm 5.4 Parameter to take into account the overlapping for the multisplitting method

---

if MyRank=0 or MyRank=NbProcs−1 **then**
    Size ← Size+Overlap
else
    Size ← Size+2×Overlap
end if
if MyRank≠0 **then**
    Offset ← Offset−Overlap
end if

---

form of the matrix can be re-used for the next iterations. So, when the number of iterations to reach the convergence threshold is high, the time to factorize a matrix may not be so important in comparison to the number of times that the factorized form will be used. If an iterative method is used, the time to solve a subsystem may vary linearly with the size of a sparse matrix. Hence, it may be worth overlapping some components but it is difficult to define an optimal overlapping size. Furthermore, the optimal size may depend on the network speed, because if the bandwidth is low, it may be preferable to compute longer and communicate less.

5.6.5   Asynchronous Newton-multisplitting algorithm

In Algorithm 4.8 we have described the synchronous version of the Newton-multisplitting algorithm. In order to define the asynchronous version of that algorithm, presented in Algorithm 5.5, we can use the same variables (c.f. Algorithm 4.7), except that instead of the variable $MaxErrorMulti$ we need a boolean $Converged$ as in all other AIAC algorithms. It should be noted that in the asynchronous Newton-multisplitting algorithm only one part is asynchronous, this is the computation of the solution of the linear system obtained at each iteration of the Newton process. So, the Newton iterations are still synchronous.

In Figure 5.1, we illustrate the behavior of the algorithm. At each Newton iteration, a synchronization step is used; it is represented by a vertical line in the figure. The synchronization corresponds to the computation of the global error of the Newton process and to the diffusion of the local values of components of vector X computed on each processor. Rectangles represent iterations of the multisplitting method used to solve the linear system obtained at each Newton iteration. So, this figure clearly highlights that Newton iterations are synchronized whereas multisplitting iterations are asynchronous.

Algorithm 5.5 Asynchronous Newton-multisplitting algorithm

repeat
  if first iteration or required **then**
    Computation of the Jacobian rectangular matrix and storage of the
    respective parts into J, $JDepLeft$ and $JDepRight$
  end if
  Computation of $-F$ depending on X from components Offset to
  Offset+size-1 and storage of the result into F
  Converged $\leftarrow$ false
  repeat
    FLoc $\leftarrow$ F
    if MyRank $\neq 0$ **then**
      FLoc $\leftarrow$ FLoc$-$JDepLeft$\times$DXLeft
    end if
    if MyRank $\neq$ NbProcs-1 **then**
      FLoc $\leftarrow$ FLoc$-$JDepRight$\times$DXRight
    end if
    DX $\leftarrow$ Solve(J, FLoc)
    for i=0 to NbProcs-1 **do**
      if i $\neq$ MyRank and DependsOnMe[i] **then**
        Send(i, PartOf(DX, i))
      end if
    end for
    for i=0 to NbProcs-1 **do**
      if i $\neq$ MyRank and IDependOn[i] **then**
        if Recv(i, TLoc) **then**
          Update DXLeft or DXRight with TLoc according to processor i
        end if
      end if
    end for
    ErrorMulti$\leftarrow 0$
    for i=0 to Size-1 **do**
      ErrorMulti $\leftarrow$ max(ErrorMulti, abs(DX[i]$-$DXOld[i]))
      DXOld[i]$\leftarrow$ DX[i]
    end for
    Converged $\leftarrow$ convergence(ErrorMulti, EpsilonMulti)
  until Converged = true
  X $\leftarrow$ X+DX
  ErrorNewton$\leftarrow 0$
  for i=0 to Size-1 **do**
    ErrorNewton $\leftarrow$ max(ErrorNewton, abs(DX[i]))
  end for
  AllToAllV(X[Offset], X, Size)
  AllReduce(ErrorNewton, ErrorNewtonMax, Max)
until stopping criteria of Newton is reached
    (MaxErrorNewton $\leq$ EpsilonNewton)

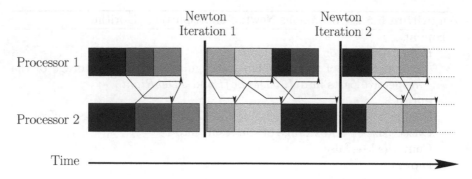

FIGURE 5.1: Iterations of the Newton-multisplitting method.

5.6.6   Asynchronous multisplitting-Newton algorithm

This algorithm allows us to solve a nonlinear system. It is formally described in Section 5.5.2. Algorithm 5.6 describes a possible implementation. In order to see the difference in terms of the computations, we represent in Figure 5.2 the decomposition of the problem. In fact, the problem is not considered in its globality like in the Newton-multisplitting method. The function F is split into *NbProcs* processors. And each of them must solve a different part of the function F using the Newton process. So, the Jacobian matrix is not considered in its totality as in the previous algorithm. Each processor computes a local Jacobian matrix of size $Size \times Size$ which corresponds to the local size after decomposition. As in the previous algorithm the computation of the local function F requires a larger part of the vector X than the one locally computed. This is why in Algorithm 5.6 we consider that this vector has the global size of the system. In opposition to the Newton-multisplitting algorithm for which the multisplitting is used to solve a linear system obtained at each Newton iteration, the multisplitting in this algorithm is used to split the nonlinear system, i.e., the Newton method. So, this method has only one iteration in which messages are used to update the approximation of the vector X that is computed locally with Newton iterations on local subsystems.

From the programming point of view, this method is simpler to implement than the Newton-multisplitting one. Figure 5.2 allows us to understand how the decomposition is different from the previous one. It is easy to see that the parts called $JDepLeft$ and $JDepRight$ in Figure 4.12 are completely ignored. Likewise, there is no need for processors to exchange their local solutions of each subsystem (in opposition to the Newton-multisplitting method) since this is done by directly exchanging vector X. As already mentioned in the previous chapter, in all the multisplitting methods it is possible to solve the subsystems using either a direct method or an iterative one. In the latter case, we obtain a two-stage algorithm.

Algorithm 5.6 Asynchronous multisplitting-Newton algorithm

NbProcs: number of processors
MyRank: rank of the processor
Size: local size of the matrix
SizeGlo: global size of the matrix
Offset: offset of the global index
JLoc[Size][Size]: local block-diagonal part of the Jacobian matrix
DependsOnMe[NbProcs]: array of the dependent processors
IDependOn[NbProcs]: array of the processors this processor depends on
F[Size]: right-hand side vector of the subsystem
X[SizeGlo]: solution vector of the subsystem
DX[Size]: solution vector of the multisplitting subsystem
TLoc[Size]: array used for the receptions of the dependencies
Error: local error
Epsilon: desired accuracy
Converged: convergence state

repeat
    if first iteration or required **then**
        Computation of the Jacobian submatrix and storage of the result into JLoc
    end if
    Computation of $-F$ depending on X from components Offset to Offset+size-1 and storage of the result into F
    DX $\leftarrow$ Solve(JLoc, F)
    for i=0 to Size-1 **do**
        X[Offset+i] $\leftarrow$ X[Offset+i]+DX[i]
    end for
    for i=0 to NbProcs-1 **do**
        if i $\neq$ MyRank and DependsOnMe[i] **then**
            Send(i, PartOf(X, i))
        end if
    end for
    for i=0 to NbProcs-1 **do**
        if i $\neq$ MyRank and IDependOn[i] **then**
            if Recv(i, TLoc) **then**
                Update X according to processor i
            end if
        end if
    end for
    Error$\leftarrow$ 0
    for i=0 to Size-1 **do**
        ErrorMulti $\leftarrow$ max(Error, abs(DX[i]))
    end for
    Converged $\leftarrow$ convergence(Error, Epsilon)
until Converged = true

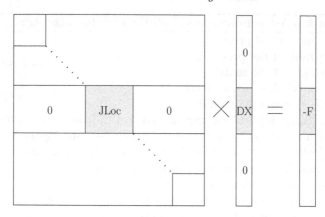

FIGURE 5.2: Decomposition of the multisplitting-Newton.

In Figure 5.3 we represent iterations of the multisplitting-Newton method. The rectangles represent the iterations of the Newton process on each processor. As can be seen, those iterations are asynchronous. So, compared to Figure 5.1 it is obvious that the multisplitting-Newton algorithm may be faster than the Newton-multisplitting one when communication delays are in favor of asynchronous iterations. This is typically the case in distant clusters in which the communication links and the machines are generally heterogeneous, implying large disparities in the communication and computation speeds in the system.

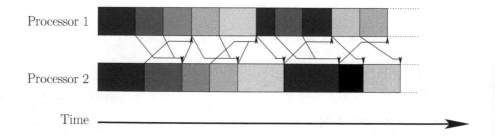

FIGURE 5.3: Iterations of the multisplitting-Newton method.

5.7   Convergence detection

As seen in Section 4.5, the convergence detection is an important issue of iterative algorithms. In the asynchronous context, the convergence detection is even hardened by the difficulty to get a correct image of the global state at any time during the process.

The most common techniques used in distributed computing to recover that information are centralized [55, 43, 100] and synchronous [84]. By their nature, those detection algorithms are efficient in parallel systems with a small physical radius but are not suited to large scale and/or distant distributed systems. Moreover, they are not suited to asynchronous iterative algorithms either, as the global synchronizations required at each recovery of the global state would indirectly synchronize the iterative process itself and then would drastically reduce the ratio of asynchronism and its benefit.

In fact, specific studies about the termination detection have been led in the context of asynchronous iterative algorithms [33, 104, 37]. But, most of them were either centralized or based on particular assumptions sometimes including some modifications of the iterative process itself.

So, in order to preserve the benefit of the asynchronism, the convergence detection algorithm must also be asynchronous. Moreover, the centralization of such an algorithm may not only generate the classical problem of bottlenecks but may also induce a loss of generality in its possible contexts of use. Indeed, in the classical centralized algorithms, all processors directly communicate their information to the central one. However, that communication scheme, implying that one machine can directly be contacted by all the others, is not possible in all parallel systems, particularly in the distributed clusters in which each site may have restricted access policies for security reasons. In most cases, only one machine of a given cluster is reachable from the outside. In order to bypass that problem, an explicit forwarding of the messages can be performed from any node in the system toward the central one. That method presents the advantage of only involving communications between neighboring nodes and is adapted to the hierarchical communication systems that can be found in distributed clusters. Unfortunately, that scheme implies more communications, slowing down the network and indirectly the iterative process itself. Moreover, it also implies larger delays toward the central node.

So, the most suitable detection algorithm in that context must not only be asynchronous but also completely decentralized. Such an algorithm is presented below.

5.7.1   Decentralized convergence detection algorithm

The decentralized algorithm for global convergence detection presented here works on all parallel iterative algorithms, either asynchronous or synchronous.

Although the version described in the following is closer to asynchronous algorithms, which represent the most general case, only a few adaptations are necessary to use it in the synchronous context.

The major difficulty with termination detection lies in the proof that the proposed algorithm does not detect convergence prematurely. Indeed, in asynchronous algorithms, the delays between iterations could lead to a false realization of the convergence criterion. This situation typically occurs in heterogeneous contexts, for example when a processor computes a new iteration whereas a slower processor computes a former iteration. That difficulty is increased with distant processors where the communication/computation ratio may be important.

As for the classical convergence detection algorithms, the principle of the decentralized detection algorithm is based on two steps. The first one consists in detecting the local convergence on each processor and the second one properly consists in the global convergence detection. Those two steps are described in the following paragraphs.

5.7.1.1 Local convergence detection

The local convergence step is quite similar to the one used in the synchronous case. As explained in Section 4.5, there is usually no information about the distance between the current state of the system and its fixed point. So, in place, the residual is used according to a chosen metric to get an idea of the stabilization of the process. Finally, that stabilization is itself determined by the setting of a threshold on the residual. However, it has also been seen in the previous chapter that when the metric used is not the contraction one, the residual does not follow a monotonous decrease but there may be oscillations around the given threshold. Hence, if no care is taken, a local convergence can be detected too early, leading in turn to a false detection of the global convergence. Once again, we insist on the fact that this problem is common to *all* iterative algorithms and is *not* due to the asynchronism.

Currently, there is no way to ensure a definitive local convergence on a processor without modifying the iterative process, as in [33]. The common heuristic is then to assume that local convergence is achieved when the node has performed a given number of successive iterations under the residual threshold. That mechanism is used in Algorithm 5.7. It implies the use of a constant, called $THRESHOLD\_LOCAL\_CV$, which represents the required number of successive iterations under the residual threshold to ensure the local convergence. It is important to note that this $THRESHOLD\_LOCAL\_CV$ value theoretically exists and is finite since, by hypothesis, the asynchronous iterative process converges. However, that value is quite difficult, not to say impossible, to evaluate in practice. Consequently, the use of an approximate value implies that the detection of the local convergence may not be definitive as the residual may rise again over the threshold after the considered number of iterations passed under it. Hence, in that context, the local state of a node may alternatively vary between convergence and non-convergence. This is why two versions of the detection algorithm are presented in the following: a theoretical version, not affected by that problem, which is useful to describe and prove the overall detection scheme, and a practical version which takes into account that problem of local states alternation.

5.7.1.2   Global convergence detection

The goal here is to obtain a similar stopping criterion as in the sequential/synchronous modes, that is to say, having all the nodes in local convergence at the same time. Unfortunately, if the asynchronism is not responsible for the difficulty in evaluating the local convergence, it hardens the global convergence detection by making the building of a representative image of the global state of the system more difficult. The process described below allows us to detect the global convergence on any one node of the system in a decentralized manner. Its correctness is proved in the context where the contraction norm is used. In other cases, often encountered in practice, the process is still correct but an additional verification step is necessary after the global detection to ensure that the system was in the correct global state at the detection time.

5.7.1.2.1   Global detection scheme: The decentralization of the detection algorithm is based upon a scheme quite similar to the leader election protocol [83]. That protocol consists in dynamically designating one processor to perform a given task. In that case, the task will be the global convergence detection. However, in that particular context, the leader election process requires some specific adaptations which imply the use of a tree graph. Fortunately, that does not reduce the generality of the algorithm since it is always possible to compute (off-line or in-line) a spanning tree from any connected graph.

The election process works with what can be called *PartialCV* messages between processors. Such a message informs the receiver that all the processors in the subtree depending on the sender (behind the sender according to the receiver) have reached local convergence. Hence, on each processor, the algorithm considers the number of neighbors (in the tree) from which no *PartialCV* message has already been received.

When that number is equal to one and the node is in local convergence, it sends a *PartialCV* message to its last neighbor which has not sent it such a message yet. It is at that point that the spanning tree is necessary. It ensures that there always exists at least one node in the system which only has one neighbor (all the leaves of the spanning tree). Thus, the partial convergence detections will propagate from the leaves of the spanning tree toward the inner nodes and will meet on one node. So, as depicted in Figure 5.4, a node will detect the global convergence when it has received the *PartialCV* messages from all its neighbors and is itself in local convergence.

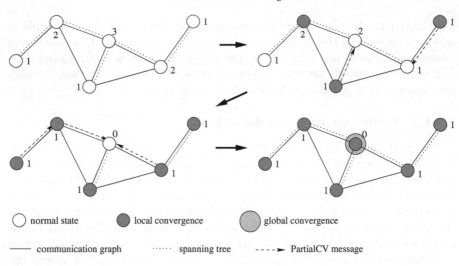

FIGURE 5.4: Decentralized global convergence detection based on the leader election protocol. For each node, the number of its neighbors in the spanning tree from which no partial convergence message has been received is indicated.

The way the process is designed implies that such a detection may happen on two neighboring nodes in place of only one. This occurs when all the nodes in the system are in local convergence and the propagation of the *PartialCV* messages ends at two neighboring nodes which are for each other the last one which has not yet sent its *PartialCV* message to the other one. So, both those nodes send their message to the other, implying a double detection of the global convergence on the two nodes. Such a particular situation is presented in Figure 5.5.

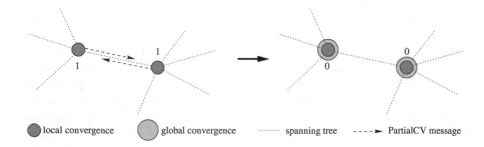

FIGURE 5.5: Simultaneous detection on two neighboring nodes.

Fortunately, that situation is not a problem per se in that context since it does not correspond to a false detection but only to a multiple one. Moreover, as the halting procedure is performed by the propagation of halting messages throughout the system from the elected node(s) and each node forwards the halting message to its other neighbors only once, that special case generates only two useless messages between the two elected nodes. So, it does not alter the halting process and does not actually require any particular treatment. However, if for some reason (often a practical one) only one node had to be elected, this could be easily achieved with, for example, a simple verification and choice mechanism between a node which detects the global convergence and its neighbor from which has come the last *PartialCV* message.

NbNeig	integer representing the number of neighbors in the spanning tree
RecvdPCV[NbNeig]	boolean array indicating for each neighbor of the current node in the spanning tree if a PartialCV message has been received from that node
NbNotRecvd	number of neighbors from which no PartialCV message has been received yet
NbUnderTh	number of successive iterations with a residual under the threshold
UnderTh	boolean equals true when the residual is under the threshold and false otherwise
LocalCV	boolean equals true when the local convergence is detected and false otherwise
GlobalCV	boolean equals true when the global convergence is detected and false otherwise

Table 5.1: Description of the variables used in Algorithm 5.7.

The decentralized detection algorithm obtained is given in Algorithm 5.7. For clarity sake, a description of the variables used in that algorithm is given in Table 5.1.

The receipts of messages are handled by distinct functions and do not directly appear in the main algorithm. That organization is particular to asynchronous algorithms where communications are not performed and managed at specific times in the algorithm but as soon as they are required or they occur.

The function *RecvPartialCV* only consists in decreasing the number of neighbors which have not yet reached local convergence. The function *recvGlobalCV* consists in stopping the iterative process on the node by setting the *GlobalCV* variable to *true*.

Algorithm 5.7 Decentralized global convergence detection

---

for all $P_i, i \in \{1, \ldots, N\}$ **do**
  NbNotRecvd $\leftarrow$ NbNeig
  for Ind from 0 to NbNeig-1 **do**
    RecvdPCV[Ind] $\leftarrow$ false
  end for
  NbUnderTh $\leftarrow$ 0
  UnderTh $\leftarrow$ false
  LocalCV $\leftarrow$ false
  GlobalCV $\leftarrow$ false
  repeat
    if LocalCV = false **then**
      ... iterative process and evaluation of UnderTh ...
      if UnderTh = true **then**
        NbUnderTh $\leftarrow$ NbUnderTh + 1
        if NbUnderTh = THRESHOLD_LOCAL_CV **then**
          LocalCV $\leftarrow$ true
        end if
      else
        NbUnderTh $\leftarrow$ 0
      end if
    end if
    if LocalCV = true **then**
      if NbNotRecvd = 0 **then**
        GlobalCV $\leftarrow$ true
      else
        if NbNotRecvd = 1 **then**
          Send a PartialCV message to the neighbor corresponding to the
          unique cell of RecvdPCV[] being false
        end if
      end if
    end if
  until GlobalCV = true
  Broadcast a GlobalCV message to all neighbors in the spanning tree from
  which no GlobalCV message has arrived
end for

Algorithm 5.8 Function RecvPartialCV()

Extract SrcNode from the message
SrcIndNeig ← corresponding index of SrcNode in the list of neighbors
                  of the current node (-1 if not in the list)
//*the test is just a precaution since such a message should always come*
//*from one of the neighbors in the spanning tree*
if SrcIndNeig ≥ 0 **then**
    RecvdPCV[SrcIndNeig] ← true
    NbNotRecvd ← NbNotRecvd-1
end if

Algorithm 5.9 Function RecvGlobalCV()

GlobalCV ← true

5.7.1.2.2 **Validity proof:** We remind the reader that the convergence detection algorithm above is to be used with any asynchronous iterative process which converges. It is important to underline that this process does not force the convergence of any asynchronous iterative process but ensures the correct convergence detection of a converging asynchronous iterative process.

Preliminary definitions:
Let $P = \{P_1, ..., P_N\}$ be the set of the processors.
Let us define NoPCVmsg(P_i, P_j, t) between two neighboring processors P_i and P_j at time t as:

$$\text{NoPCVmsg}(P_i, P_j, t) =$$
$$\begin{cases} \text{true if } P_i \text{ has not yet received a PartialCV message from } P_j \\ \text{false if } P_i \text{ has received a PartialCV message from } P_j \end{cases}$$

The detection algorithm is based on two particular properties of the processors which are the local convergence and the number of neighbors having communicated their partial convergence. Since these properties evolve during the iterative process, the set $P(t)$ of processors P_i can be written as the following partition:

$$P(t) = \quad S_0^c(t) \cup S_1^c(t) \cup \ldots \cup S_{N-1}^c(t)$$
$$\cup\, S_0^d(t) \cup S_1^d(t) \cup \ldots \cup S_{N-1}^d(t)$$

where $S_k^e(t)$ is the set of processors having at time t:

$$NbNotRecvd = k$$
$$LocalCV = \begin{cases} \text{true if } e = c \\ \text{false if } e = d \end{cases}$$

The particular presentation of $P(t)$ is only for intuitive representation of the partition.

Finally, we note $t_c(i)$ the time at which processor P_i reaches local convergence and we define $t_r(k, j)$ as the receipt time of the *PartialCV* message on P_k from P_j and $t_m(j, k, t)$ as the communication time from P_j to P_k at time t (t is included because communication times may vary during the process). We have then:

$$t_r(k, j) = t_c(j) + t_m(j, k, t_c(j))$$

THEOREM 5.7

If the following hypotheses are satisfied:

(H1) The communication graph used for the detection process is connected and acyclic

(H2) The asynchronous iterative process converges

(H3) Communications between neighbors are achieved in a finite time

then, there exists $t_d \in \mathbb{N}$ such that

$$\begin{aligned}
&S_0^c(t_d) \neq \emptyset \\
&|S_1^c(t_d)| \geq 0 \\
&S_k^c(t_d) = \emptyset \quad k \in \{2, ..., N-1\} \\
&S_k^d(t_d) = \emptyset \quad k \in \{0, ..., N-1\}
\end{aligned}$$

$\square$

The second statement only appears to point out that there is no particular condition on $S_1^c(t_d)$.

The proof of Theorem 5.7 is made in two steps:

(A) we prove that $S_0^c(t_d) \neq \emptyset$ implies all the other statements of Theorem 5.7

(B) we prove that $\exists t_d \in \mathbb{N}$ such that $S_0^c(t_d) \neq \emptyset$

Part (A):

Let us define $\text{Neigh}(P_i)$ the set of physical neighbors of processor P_i. In order to get the processor P_i in $S_0^c(t)$, we must have by Algorithm 5.7:

$$\forall P_j \in \text{Neigh}(P_i), \; \text{NoPCVmsg}(P_i, P_j, t) = false$$

which implies in turn for all the P_j that

$$\forall P_k \in \text{Neigh}(P_j) \setminus \{P_i\}, \; \text{NoPCVmsg}(P_j, P_k, t) = false$$

and by recursion, we deduce that

$$\forall P_a \in P(t) \setminus \{P_i\}, \; \exists P_b \in P(t), \; \text{NoPCVmsg}(P_b, P_a, t) = false \qquad (5.42)$$

This means that all the P_a in that equation have sent a *PartialCV* message to the corresponding P_b and by Algorithm 5.7, this is only possible once P_a has reached local convergence.

Thus, we have:

$$\forall P_a \in P(t) \setminus \{P_i\}, \ P_a \notin \bigcup_{u=0}^{N-1} S_u^d(t)$$

and since $P_i \in S_0^c(t)$, then

$$\bigcup_{u=0}^{N-1} S_u^d(t) = \emptyset$$

Moreover, by Algorithm 5.7, we also know that the condition for a processor P_a to verify Equation (5.42) (sending of a *PartialCV* message to another node) is to have its *NbNotRecvd* equal to one.

Hence:

$$\forall P_a \in P(t) \setminus \{P_i\}, \ P_a \in \bigcup_{u=0}^{1} S_u^c(t)$$

and then

$$\bigcup_{u=2}^{N-1} S_u^c(t) = \emptyset$$

and all the other statements of Theorem 5.7 are verified. $\square$

Part (B):

By definition, at the beginning of the process, the following statements are verified:

$$\begin{aligned} S_k^c(0) &= \emptyset \quad \forall k \in \{0, ..., N-1\} \\ S_0^d(0) &= \emptyset \\ S_1^d(0) &\neq \emptyset \end{aligned} \tag{5.43}$$

The third statement comes from (H1) which implies that the graph always has at least one node with only one neighbor.

By (H2), we have:

$$\begin{aligned} P_i \in S_k^d(t), \ i \in \{1, ..., N\}, \ k \in \{0, ..., N-1\} \\ \Rightarrow \exists \, t_c(i) \in \mathbb{N}, \ \forall t \geq t_c(i), P_i \in \bigcup_{u=0}^{k} S_u^c(t) \end{aligned} \tag{5.44}$$

hence

$$\exists t'(k) \in \mathbb{N}, \forall t \geq t'(k), \ |S_k^d(t)| = 0, \ k \in \{0, ..., N-1\} \tag{5.45}$$

Equation (5.43), Equation (5.45) and Algorithm 5.7 imply that

$$\exists t_{dn}, \quad \begin{cases} S_1^c(t_{dn}) \neq \emptyset \\ \forall t < t_{dn}, \ \bigcup_{u=0}^{N-1} S_u^d(t) \neq \emptyset \\ \forall t \geq t_{dn}, \ \bigcup_{u=0}^{N-1} S_u^d(t) = \emptyset \\ \forall t < t_{dn}, \ S_0^c(t) = \emptyset \end{cases} \tag{5.46}$$

The last statement is, in fact, a deduction from the second one. As seen in part (A), $S_0^c(t) \neq \emptyset$ implies that $\bigcup_{u=0}^{N-1} S_u^d(t) = \emptyset$ which is in contradiction with the second statement for each $t < t_{dn}$.

Now, at t_{dn}, we know by Equation (5.46) that $S_1^c(t_{dn}) \neq \emptyset$. So, every $P_i \in S_1^c(t_{dn})$, according to Algorithm 5.7, sends a *PartialCV* message to its unique neighbor P_k which verifies NoPCVmsg$(P_i, P_k, t_{dn}) =$ true. We define:

$$A(t) = \{ P_i \in S_1^c(t), \ \exists! P_k \in P(t),$$
$$\text{NOpCVmess}(P_i, P_k, t) = \text{NOpCVmess}(P_k, P_i, t) = \text{true} \}$$

and

$$B(t) = \{ P_k \in P(t), \ \exists P_i \in A(t) \ \text{such that NoPCVmsg}(P_i, P_k, t) = \text{true} \}$$

So, $A(t)$ is the set of processors whose sending of the *PartialCV* message to exactly one element of $B(t)$ (corresponding set of destination nodes) has not yet arrived at time t.

From (H1), we deduce the following lemma.

LEMMA 5.1
Considering the set A and time $t' \geq t_{dn}$:

$$A(t'-1) \neq \emptyset, \ A(t') = \emptyset \quad \Rightarrow \quad \begin{cases} \forall t \geq t', A(t) = \emptyset \\ \exists P_i \in S_0^c(t') \end{cases}$$

$\square$

Justification of Lemma 5.1:

Since $t' \geq t_{dn}$, we are in the context of Equation (5.46) where all the processors are in the subsets $S_u^c, u \in \{0, ..., N-1\}$.

If we consider the state of the system at time t', it is not possible to have one node in another subset than S_0^c or S_1^c since this would imply that this node has not yet received the *PartialCV* message from at least two of its neighbors.

So, either these neighbors are communicating their *PartialCV* message to this node, which is a contradiction to $A(t') = \emptyset$, or the other possibility is that these neighbors have not sent their *PartialCV* message to this node yet. Nevertheless, the only way for these neighbors not to have sent their *PartialCV* message to this node yet is that they have themselves at least two neighbors from which they have not received the *PartialCV* message yet. If we continue this reasoning by recursion, we come to the conclusion that this situation is only possible if all these nodes form a cycle in the graph which is a contradiction to hypothesis (H1).

Hence, we are sure that all the nodes have reached their local convergence and sent a *PartialCV* message which has already arrived at the destination node.

Finally, (H1) also implies that there is at least one node which has received the *PartialCV* messages from all its neighbors and is then located in $S_0^c(t_{dn})$.

REMARK 5.1   One consequence is that as soon as the set A becomes empty, it cannot become nonempty again. ⬚

REMARK 5.2   Another consequence is that time t' is equivalent to time t_d in Theorem 5.7 since $S_0^c(t') \neq \emptyset$ and then Part (A) of the proof implies all the other statements of the theorem. ⬚

REMARK 5.3   At time t_{dn}, all the processors have reached their local convergence and since $S_1^d(0) \neq \emptyset$ it is sure that the set A becomes nonempty at the latest at time t_{dn}. ⬚

Now, let us examine the set $A(t_{dn})$:

If it is empty, Lemma 1 and Remark 5.3 imply that it was nonempty at the time just before and then t_{dn} corresponds to the time t' in Lemma 1 which also corresponds to the time t_d in Theorem 5.7 as pointed out by Remark 5.2.

If it is nonempty, Equation (5.46) implies that $B(t_{dn}) \subseteq \bigcup_{u=1}^{N-1} S_u^c(t_{dn})$ and there are two distinct possibilities over the set $B(t_{dn})$: (5.47)

(1) $\forall P_l \in B(t_{dn}), P_l \in S_1^c(t_{dn})$

(2) $\forall P_l \in B(t_{dn}), P_l \in \bigcup_{u=2}^{N-1} S_u^c(t_{dn})$

Case (1):

In this case, there exists at least one $P_l \in B(t_{dn})$ such that $\exists! P_i \in A(t_{dn})$ for which $\text{NoPCVmsg}(P_l, P_i, t_{dn}) = \text{true}$ and $\text{NoPCVmsg}(P_l, P_i, t_r(l, i)) = \text{false}$ implying $P_l \in S_0^c(t_r(l, i))$, and leading to the detection of the global convergence on P_l at time $t_r(l, i)$. Hypothesis (H3) ensures that $t_r(l, i) < \infty$ and then statement (B) is verified with $t_d = t_r(l, i)$.

Case (2):

In this case, $P_l \in B(t_{dn})$ implies that there is one $S_u^c(t_{dn})$, $u \in \{2, ..., N-1\}$ such that $P_l \in S_u^c(t_{dn})$, and then by Algorithm 5.7:

$$P_l \in \bigcup_{v=0}^{u-1} S_v^c(t_r(l, i)), \quad \text{with } i \text{ such that } P_i \in A(t_{dn}) \tag{5.48}$$

$$\text{and NoPCVmsg}(P_i, P_l, t_{dn}) = \text{true}$$

This means that each time a processor receives a *PartialCV* message, its number of neighbors which have not sent it a *PartialCV* message yet decreases by one. Moreover, we use a union of the $u - 1$ first subsets because this processor may receive other *PartialCV* messages from other neighbors in the interval time between t_{dn} and $t_r(l, i)$, making it move down by more than one subset.

Hence, by Lemma 1:

- either there exists at least one $P_l \in B(t_{dn})$ for which $P_l \in A(t_r(l, i))$, with $t_r(l, i) < \infty$ by (H3), and we come back to a similar context as in (5.47) where $A(t) \neq \emptyset$ by replacing t_{dn} by $t_r(l, i)$ and we obtain a recursion on $\bigcup_{u=1}^{N-1} S_u^c(t)$. Equation (5.48) ensures that this recursion will empty all the subsets $S_u^c(t)$, $u \in \{2, ..., N-1\}$ and will then converge toward case (1).

- or none of the nodes of $B(t_{dn})$ comes in the set A which becomes empty as soon as all the nodes of $B(t_{dn})$ have received their *PartialCV* message (in a finite time by (H3)), directly leading to Theorem 5.7 by Remark 5.2.

$\square$

As a last remark, it can be noticed that hypothesis (H3) also implies that the termination of the iterative process on all nodes happens in a finite time after the global convergence detection on the elected node (at time t_d in Theorem 5.7).

5.7.1.2.3 Practical version: As mentioned at the beginning of Section 5.7.1.2, Algorithm 5.7 is only usable in that form when the contraction metric is known and used to compute the residual. When that metric is unknown, the difficulty of ensuring the local convergence on each node implies the use of two additional mechanisms: an optional one which is useful to regulate the local detections better, and a vital one which permits us to get a correct image of the global state of the system. Indeed, the possible alternation of the local state of the nodes requires a more accurate snapshot of the global state of the system to ensure that all the nodes have verified the local convergence conditions at the same time.

The practical version presented here is somewhat different from the one proposed in [17]. That previous version has the drawback of requiring the determination of the maximal communication time between any couple of nodes in the system during the entire iterative process. In practice, it is quite difficult, if not to say impossible, to get an accurate estimation of such a value. The version described here does not use that value and, more generally, presents the advantage of not requiring any specific information on the parallel system used. Its approach is closer to the theoretical version presented in the previous part in the sense that it lets the global detection happen even if the local evolutions on the nodes change during the election process. Then, after the global detection, it includes an additional verification phase to ensure its validity. It is important to notice here that the iterative process is not interrupted either during the global convergence detection process or during the verification phase. There are two reasons for that; the most obvious one is not to slow down the iterative process itself and the second one is that its evolution during the global detection and verification processes represents a mandatory piece of information.

Concerning the first of the two mechanisms mentioned above, it concerns the local convergence detection on each node and consists in taking into account what we call pseudo-periods in place of a given number (arbitrary in practice) of successive iterations. In the domain of dynamic systems, a period corresponds to a minimal span of time during which all the components of the system are updated at least once with different data values from its dependencies. The pseudo-period is quite a local version of that global progression step. So, for each node, a pseudo-period corresponds to the minimal span of time during which that node receives at least one newer data message from all its dependencies. In this way, the local evolution of one node is fully representative between two consecutive pseudo-periods. Thus, the local convergence detection is no longer assumed after a given number of successive iterations with the residual under the threshold but after at least one (but possibly several successive ones) pseudo-period verifying that constraint. This has a drastic regulating effect on the local convergence detections in practice and, if it cannot avoid all the false detections due to an inadequate used norm, it sharply limits them. It is therefore strongly recommended although not essential.

For its part, the second mechanism is imperative and takes place at the global level of the system just after the global convergence detection. Its aim is to verify that all the nodes were still in local convergence at the time of the global detection and that their states were representative of their evolution. Hence, that verification is decomposed in four main steps:

1) Diffusion of a verification message from the elected node through the spanning tree to initiate the verification phase.

2) Elaboration on each node of its response to the verification request.

3) Gathering of the responses of all the nodes toward the elected node through the spanning tree to get the verdict.

4) Diffusion of a verdict message from the elected node through the spanning tree to finish the verification phase.

When one node is elected by the global convergence detection process, it sends a verification message to all its neighbors (step 1). Each node which receives such a message from one of its neighbors (referred to as the asking node in the following) forwards it to all its other neighbors in the spanning tree (step 1) and, while waiting for their responses, elaborates its own response (step 2). The response of a node does not only depend on its own state and evolution but also on the responses of its neighbors in the spanning tree, except the asking one. As soon as the response is available, the node returns it to its asking node (step 3). Finally, when the elected node has its own response and those of its neighbors, it deduces the verdict and sends it to all its neighbors (step 4). Then, each node receiving a verdict message forwards it to its other neighbors in the spanning tree (step 4). At the end of the verification phase, the state of each node is set up according to the final verdict. The global scheme of that verification mechanism is depicted in Figure 5.6.

As mentioned above, the response of each node depends on its state but also on its evolution during the verification phase. Effectively, in order to ensure that all the nodes have been in local convergence at the same time (which is the criterion used in the sequential and synchronous versions), the response of a node is positive if and only if its residual never goes back over the threshold during the span of time between its last sending of a *PartialCV* message and the sending of its response to the verification request.

Moreover, to be sure that the response of each node is representative of its actual state and is not illusory, a particular mechanism is inserted to ensure that each node actually evolves during the span of time between the receipt of the verification request and the sending of its response. That mechanism roughly corresponds to the waiting of a particular pseudo-period. The ideal way to ensure the pertinence of the global state image would be to wait for a period and watch the resulting state. However, periods are quite difficult and expensive to identify in dynamical systems implemented on distributed environments. So, a lighter concept is used here which is better suited to the decentralization constraint while giving pertinent information about the evolution of the system as well. Hence, each node sends its response (depending on its residual evolution) only after having performed at least one iteration with versions of all its data dependencies at least as recent as the global detection time. In this way, the response will be fully representative of the actual evolution and state of that node until that time.

In order to force the nodes to use specific data versions during the verification phase, a tagging system is included in the data messages in order to differentiate them between the successive phases of the iterative process (normal processing and verification phase). Moreover, since there may be several verification phases during the whole iterative process, due to possible cancellations of global detections, that tagging is also useful to distinguish the data

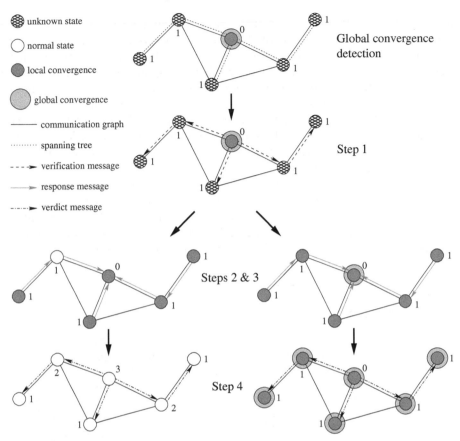

FIGURE 5.6: Verification mechanism of the global convergence. The two possibilities are illustrated, cancellation on the left and confirmation on the right. The unknown states in the first two steps are from the elected node point of view. The value beside each node corresponds to the variable *NbNotRecvd*.

messages related to the different verification phases. Finally, there is another good reason to use tagged messages not only for the data communications but also for the information related to the global detection and verification processes: the reactivity of the verification phase.

In fact, as quite an important number of verification phases is likely to occur during an entire iterative process, it is rather important to increase its reactivity. This has the indirect effect of reducing the latency between the actual global convergence of the system and its detection. In order to do so, each node is allowed to send its response as soon as it is able to deduce it, that is to say, when its residual goes over the threshold or when it receives a negative response from one of its neighbors. Such events imply a negative response of the node, whatever the values of the other elements constituting

its response are. It is then a waste of time to wait for the responses of the
other neighbors or the local completion of a pseudo-period. The same behavior
takes place on the elected node except that it directly sends a negative verdict
message to its neighbors instead of sending a response to an asking node.
However, as the order of the messages is not ensured in the asynchronous
computing context, this strategy also implies that messages related to a given
verification phase may arrive on a node after the termination of that phase
(in the case of a verification phase canceled faster). Thus, in order to avoid
confusions in the messages related to the global detection scheme, a tagging
system must also be inserted.

Finally, in order to respond to all those message distinction constraints,
each phase of the iterative process (normal computing and verification of the
global convergence) is distinguished in time by an integer tag incremented
at each phase transition, as shown in Figure 5.7, with four nodes linearly
organized for a span of time beginning with the tag equal to k.

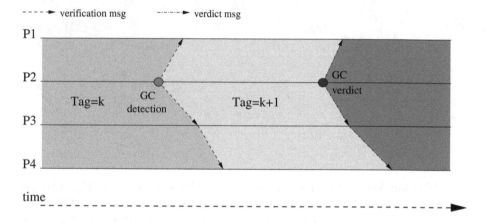

FIGURE 5.7:   Distinction of the successive phases during the iterative
process.

The whole detection and verification mechanism is detailed in Figure 5.8 in
the same computing context as above and in the case of a global convergence
detected and confirmed on node P_2. As can be seen, the whole process ensures,
in case of a positive verdict, that all the nodes in the distributed system
have had their residual under the threshold at least at the time at which the
global convergence was detected on the elected node, and possibly during a
larger span of time after it. Moreover, the pseudo-period performed on each
node during the verification phase with data as recent as the global detection
ensures that their states are representative of their actual evolutions. In this
way, that entire global convergence detection mechanism provides a similar

stopping criterion as in the sequential/synchronous cases.

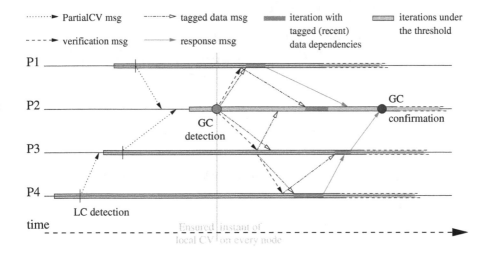

FIGURE 5.8: Mechanism ensuring that all the nodes are in representative stabilization at least at the time of global convergence detection.

As the behavior of the nodes is not the same according to the different steps in the detection process and verification phase, it is also necessary to introduce four different states:

- **NORMAL:** the basic state during the whole iterative process when the node is not in the global convergence detection mechanism.

- **WAIT4V:** when the node is waiting for the local start of the verification phase after its sending of a PartialCV message.

- **VERIF:** when the node is performing the verification phase, either after the receipt of the corresponding message or by election.

- **FINISHED:** when the global convergence has been confirmed.

The transitions between those states are depicted in Figure 5.9.

The final scheme obtained is given in Algorithm 5.15. In order to get an easy reading of it, the list of the additional variables according to the previous algorithm is given in Table 5.2.

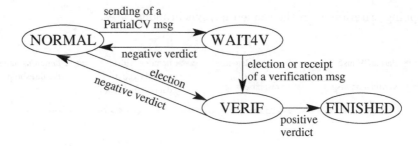

FIGURE 5.9: State transitions in the global convergence detection process.

The different types of messages are listed below together with their contents:

- **data message:**
    - identifier of the source node
    - source node iteration number at the sending time
    - source node phase tag at the sending time
    - data

- **PartialCV message:**
    - identifier of the source node
    - source node phase tag at the sending time

- **verification message:**
    - identifier of the source node
    - source node phase tag at the sending time

- **response message:**
    - identifier of the source node
    - source node phase tag at the sending time
    - response of the source node

- **verdict message:**
    - identifier of the source node
    - new phase tag to use on the receiver
    - verdict

The algorithm also uses additional functions which are briefly described below:

- **InitializeState():** (re-)initializes the variables related to the convergence detection process and sets the node in NORMAL state.

- **ReinitializePPer():** (re-)initializes the variables related to the pseudo-period detection.

MyRank	integer identifying uniquely the current node
State	integer indicating the current state of the node among NORMAL, WAIT4V, VERIF and FINISHED
PhaseTag	integer identifying the current phase on the current node
PseudoPerBeg	boolean indicating that a pseudo-period has begun
PseudoPerEnd	boolean indicating the end of a pseudo-period
NbDep	integer representing the number of computational dependencies of the current node
NewerDep[NbDep]	boolean array indicating for each data dependency if a newer version has been received since the last pseudo-period
LastIter[NbDep]	integer array indicating for each dependency node the iteration of production of the last data received from that node
PartialCVSent	boolean indicating that a PartialCV message has been sent
ElectedNode	boolean indicating that the node is the elected one
Resps[NbNeig]	integer array containing the responses of the neighbors of the current node in the spanning tree. The values are either −1 (negative), 0 (no response yet) or 1 (positive)
ResponseSent	boolean indicating that the response has been sent

Table 5.2: Description of the additional variables used in Algorithm 5.15.

- **InitializeVerif():** initializes the verification phase. In particular, the *PhaseTag* variable is incremented to distinguish the new verification phase from the potential previous ones.

- **RecvDataDependency():** manages the receipts of data dependencies. In the general asynchronous model, each received datum is taken into account, whenever it was produced. However, taking only newer data (produced after the locally available ones) tends in practice to speed up the iterative process. So, the function takes into account any newer data when the receiver is not in verification phase (VERIF state). Otherwise, it filters the data produced after the last global convergence detection, that is to say, with the same phase tag as the receiver.

- **RecvPartialCV():** manages the receipts of PartialCV messages. Also updates the local state of the node when an election is possible. However, the mutual exclusion mechanism mentioned on page 149 is performed to ensure that only one node is elected in the system.

- **RecvVerification():** manages the receipts of verification messages. The message is taken into account only when its phase tag corresponds to the following phase on the receiver. In that case, the state of the receiver is changed to enter the verification phase (VERIF state) and the message is propagated to its other neighbors in the spanning tree.

- **RecvResponse():** manages the receipts of response messages. The message is taken into account only when the phase tag in the message corresponds to the current phase tag on the receiver.

- **RecvVerdict():** manages the receipt of the verdict of the verification phase on the non-elected nodes. The verdict is always taken into account and propagated through the spanning tree to set all the nodes either in FINISHED state or back in NORMAL state with a new phase tag. As the state of the non-elected nodes cannot change before the receipt of that message, no other global convergence detection may happen before all the nodes have received it. Therefore, there cannot be any confusion with a similar message coming from a previous verification phase.

- **ChooseLeader(integer, integer):** takes two integer parameters identifying two nodes which are potential candidates to the leader election and returns the one which is chosen by the election referee policy.

The last function of the list is not detailed in the following since it directly depends on the referee policy used. The choice of that policy is quite free as its only constraint is to make a choice between the two proposed nodes.

Algorithm 5.10 Function InitializeState()

NbNotRecvd ← NbNeig
for Ind from 0 to NbNeig−1 **do**
    RecvdPCV[Ind] ← false
end for
ElectedNode ← false
LocalCV ← false
PartialCVSent ← false
ReinitializePPer()
State ← NORMAL

Algorithm 5.11 Function ReinitializePPer()

PseudoPerBeg ← false
PseudoPerEnd ← false
for Ind from 0 to NbDep−1 **do**
    NewerDep[Ind] ← false
end for

Algorithm 5.12 Function InitializeVerif()

ReinitializePPer()
PhaseTag ← PhaseTag + 1
for Ind from 0 to NbNeig−1 **do**
  Resps[Ind] ← 0
end for
ResponseSent ← false

Algorithm 5.13 Function RecvDataDependency()

Extract SrcNode, SrcIter and SrcTag from the message
SrcIndDep ← corresponding index of SrcNode in the list of dependencies
      of the current node (−1 if not in the list)
if SrcIndDep ≥ 0 **then**
  if LastIter[SrcIndDep] < SrcIter
    and (State ≠ VERIF **or** SrcTag = PhaseTag) **then**
    Put the data in the message at their corresponding place according to
    SrcIndDep in the local data array used for the computations
    LastIter[SrcIndDep] ← SrcIter
    NewerDep[SrcIndDep] ← true
  end if
end if

Algorithm 5.14 Function RecvPartialCV()

Extract SrcNode and SrcTag from the message
SrcIndNeig ← corresponding index of SrcNode in the list of neighbors
      of the current node (−1 if not in the list)
if SrcIndNeig ≥ 0 **and** SrcTag = PhaseTag **then**
  RecvdPCV[SrcIndNeig] ← true
  NbNotRecvd ← NbNotRecvd−1
  if NbNotRecvd = 0 **and** PartialCVSent = true
    and ChooseLeader(MyRank, SrcNode) = MyRank **then**
    ElectedNode ← true
    InitializeVerif()
    Broadcast a verification message to all its neighbors
    State ← VERIF
  end if
end if

Algorithm 5.15 Practical version of Algorithm 5.7   (1/3)

---

for all $P_i, i \in \{1, \ldots, N\}$ **do**
   InitializeState()
   UnderTh ← false
   PhaseTag ← 0
   repeat
     ... iterative process, data sendings and evaluation of UnderTh ...
     if State = NORMAL **then**
       if UnderTh = false **then**
         ReinitializePPer()
       else
         if PseudoPerBeg = false **then**
           PseudoPerBeg ← true
         else
           if PseudoPerEnd = true **then**
             LocalCV ← true
             if NbNotRecvd = 0 **then**
               ElectedNode ← true
               InitializeVerif()
               Broadcast a verification message to all its neighbors
               State ← VERIF
             else
               if NbNotRecvd = 1 **then**
                 Send a PartialCV message to the neighbor corresponding
                 to the unique cell of RecvdPCV[] being false
                 PartialCVSent ← true
                 State ← WAIT4V
               end if
             end if
           else
             if all the cells of NewerDep[] are true **then**
               PseudoPerEnd ← true
             end if
           end if
         end if
       end if
     else if State = WAIT4V **then**
       see that part on page 167...
     else if State = VERIF **then**
       see that part on pages 167 and 168...
     end if
   until State = FINISHED
end for

Algorithm 5.15 bis Practical version of Algorithm 5.7   (2/3)

---

for all $P_i, i \in \{1, \ldots, N\}$ **do**

  see that part on page 166...

  repeat

    ... iterative process, data sendings and evaluation of UnderTh ...

    if State = NORMAL **then**

      see that part on page 166...

    else if State = WAIT4V **then**

      if UnderTh = false **then**

        LocalCV ← false

      end if

    else if State = VERIF **then**

      if ElectedNode = true **then**

        if UnderTh = false **or** LocalCV = false

        or at least one cell of Resps[] is negative **then**

        PhaseTag ← PhaseTag + 1

        Broadcast a negative verdict message to all its neighbors

        InitializeState()

      else

        if PseudoPerEnd = true **then**

          if there are no more 0 in Resps[] **then**

            if all the cells of Resps[] are positive **then**

              Broadcast a positive verdict message to all its neighbors

              State ← FINISHED

            else

              PhaseTag ← PhaseTag + 1

              Broadcast a negative verdict message to all its neighbors

              InitializeState()

            end if

          end if

        else

          if all the cells of NewerDep[] are true **then**

            PseudoPerEnd ← true

          end if

        end if

      end if

      else

        see that part on page 168...

      end if

    end if

  until State = FINISHED

end for

Algorithm 5.15 ter Practical version of Algorithm 5.7 (3/3)

for all $P_i, i \in \{1, \ldots, N\}$ do
  see that part on page 166...
  repeat
    ... iterative process, data sendings and evaluation of UnderTh ...
    if State = NORMAL then
      see that part on page 166...
    else if State = WAIT4V then
      see that part on page 167...
    else if State = VERIF then
      if ElectedNode = true then
        see that part on page 167...
      else
        if ResponseSent = false then
          if UnderTh = false or LocalCV = false
          or at least one cell of Resps[] is negative then
          Send a negative response to the asking neighbor
          //by construction, that is the neighbor to which has been sent
          //the last PartialCV message ⇔ false cell of RecvdPCV[]
          ResponseSent ← true
        else
          if PseudoPerEnd = true then
            if there remains only one 0 in Resps[] then
            //that last 0 is located in the cell of the asking neighbor
            if the other cells of Resps[] are all positive then
              Send a positive response to the asking neighbor
            else
              Send a negative response to the asking neighbor
            end if
            ResponseSent ← true
           end if
          else
            if all the cells of NewerDep[] are true then
              PseudoPerEnd ← true
            end if
           end if
          end if
        end if
      end if
    end if
  until State = FINISHED
end for

Algorithm 5.16 Function RecvVerification()

Extract SrcNode and SrcTag from the message
if SrcTag = PhaseTag + 1 **then**
  InitializeVerif()
  State ← VERIF
  Broadcast the verification message to all its neighbors but SrcNode
end if

Algorithm 5.17 Function RecvResponse()

Extract SrcNode, SrcTag and SrcResp from the message
SrcIndNeig ← corresponding index of SrcNode in the list of neighbors
          of the current node (−1 if not in the list)
if SrcIndNeig ≥ 0 **and** PhaseTag = SrcTag **then**
  Resps[SrcIndNeig] ← SrcResp
end if

Algorithm 5.18 Function RecvVerdict()

Extract SrcNode, SrcTag and SrcVerdict from the message
if SrcVerdict is positive **then**
  State ← FINISHED
else
  InitializeState()
  PhaseTag ← SrcTag
end if
Broadcast the verdict message to all its neighbors but SrcNode

5.8 Exercises

1. Consider a linear system

$$Ax = b,$$

where A is an M-matrix and ε is a positive scalar. Prove that asynchronous algorithms associated to the Jacobi algorithm for solving this linear system converge.

2. Consider a linear system

$$(A + \varepsilon)\, x = b,$$

where A is an M-matrix. Prove that for any $\varepsilon > 0$, asynchronous algorithms associated to the Jacobi algorithm for solving this linear system converge.

3. Consider a linear system $Ax = b$. Prove that iterative algorithms arising from the splitting of A converge asynchronously if A is monotone and the splittings are weak regular.

4. Give examples of large linear systems involving monotone matrices A and by considering weak regular splitting of A, write programs of parallel multisplitting iterative algorithms that converge asynchronously to the solution of $Ax = b$ (b given).

5. Consider the gradient iterations

$$x^{(k+1)} = x^{(k)} - \gamma Ax^{(k)}.$$

Prove that if A is diagonally dominant for a sufficiently small γ, then the gradient algorithm converges asynchronously to the solution of

$$\min \frac{1}{2}x^T A x.$$

6. Consider the 2 dimensional Dirichlet problem

$$-\Delta u = f \text{ on } \Omega = \,]0,1[\, \times \,]0,1[$$
$$u = 0 \text{ on the boundary } \partial\Omega\, of\, \Omega.$$

(a) By using the finite difference method to approximate the second derivatives and following the illustration example of Chapter 1, show that the approximate solution is the solution of a linear system

$$Ax = b$$

where A is a block tridiagonal and where each block is also a tridiagonal matrix of the form

$$\begin{pmatrix} 4 & -1 & & \\ -1 & \ddots & \ddots & \\ & \ddots & \ddots & -1 \\ & & -1 & 4 \end{pmatrix}.$$

(b) Propose a convergent asynchronous algorithm to solve the approximate solution of the Dirichlet problem.

(c) Propose a parallel block convergent asynchronous algorithm to solve the approximate solution of the Dirichlet problem.

7. Consider the ordinary differential equation

$$\begin{cases} \frac{dx}{dt} = f(x,t) \\ x(0) = x_0 \\ t \in [0,T]. \end{cases} \tag{5.49}$$

Let's denote by C^1 the space of continuous functions defined on $[0,T]$ with values in $\mathbb{R}^n$. Then, we suppose that the unknown function $x \in C^1$ and that f is a continuous function.

(a) Prove that the following mapping is a norm on C^1, for $x = (x_1, ..., x_n)$,

$$n(x) = \|x\|_\infty = \max_{1 \leq i \leq n} \max_{0 \leq t \leq T} \|x_i(t)\|$$

(b) We suppose that f is Lipschitz continuous with respect to x, with constant L, i.e.,

$$\|f(x,t) - f(y,t)\|_\infty \leq L \|x - y\|_\infty$$

and consider the following fixed point mapping

$$T(x) = y \Leftrightarrow \frac{dy_i}{dt} = f(x_1, ..., x_n, t), \ y_i(0) = (x_0)_i.$$

Let K be a real number such that

$$\frac{1 - e^{-KT}}{K} < \frac{1}{L}$$

Prove that T is contractive with respect to the norm $\|x\|_K = \max_{1 \leq i \leq n} \max_{0 \leq t \leq T} e^{-Kt} \|x_i(t)\|$.

(c) Propose a parallel asynchronous algorithm which converges to the solution of (5.49).

(d) Following Section 5.4.3 of Chapter 5, build a parallel asynchronous multisplitting algorithm to solve (5.49).

8. Implement a centralized detection convergence procedure and a decentralized one. Then compare the performances on algorithms presented in this chapter.

9. With AIAC multisplitting algorithms (linear or not), compare the behavior of versions using different solvers for solving linear subsystems. It is interesting to compare the behavior of iterative solvers against direct ones. Nevertheless, the comparison between a simple iterative solver and a more complex one, like GMRES, is also instructive. Try to point out cases where simple iterative solvers perform faster than more complex ones. Try to explain when this situation is possible.

10. Try to implement the tips described in Section 5.6.4 which consist in waiting for some new messages to arrive before running the next iterations with multisplitting algorithms for solving linear systems. According to the number of neighbors of each processor try to define an appropriate number of messages to wait for before running the next iterations.

11. Try to implement the same mechanisms as in the previous exercise with nonlinear multisplitting algorithms.

12. Compare the behavior of the Newton-multisplitting algorithm and the multisplitting-Newton algorithm for some nonlinear problems. Try to point out the threshold for which one of those algorithms seems better than the other in a distant cluster context.

Chapter 6

Programming Environments and Experimental Results

Introduction

In the two previous chapters, synchronous and asynchronous algorithms for solving common problems have been described formally and their implementations have been explained. To implement those algorithms, it is possible to use programming environments which are not dedicated to the implementation of asynchronous algorithms. However, it is obviously preferable and easier to implement them using a dedicated environment. In order to implement a synchronous algorithm, any message passing environment can be used. As mentioned previously, the implementation of an asynchronous algorithm requires the disassociation of the computations from the communications. That is why in this chapter we first present some environments that we have used to implement asynchronous algorithms. Those environments are: Corba OmniOrb 4, PM2 and MPICH/Madeleine which is a multithreaded implementation of MPI. With them, the programmer needs to explicitly manage the features of AIAC algorithms. Later, two programming and execution environments have been designed to implement AIAC algorithms, namely JACE and CRAC. We present them and we explain what interesting features they provide to implement AIAC algorithms.

Finally, we report in this chapter some experiments that we have performed in different contexts of execution with different algorithms. Our goal is not to focus on pure performances that actually depend on the execution context but rather on the differences between the behaviors of the synchronous and asynchronous versions of the same algorithm. As we could not experiment with all the algorithms described in this book in the same experimental contexts, we report some experiments for some algorithms and analyze the results. From those experiments, it is difficult to deduce general rules and to say a priori, for a given problem in a given computing context, which of the synchronous or the asynchronous versions of an algorithm will be the faster. Nevertheless, the conducted experiments allow us to conclude that a crucial point is the ratio between the computation time and the communication time. That is why a brief discussion about that ratio is placed before the presentation of the experiments.

6.1 Implementation of AIAC algorithms with non-dedicated environments

In [16] we have compared three environments for implementing AIAC algorithms. These three environments are PM2 [89], MPICH/Madeleine [7] and OmniOrb 4 [1].

The first one was naturally selected since we implemented our first AIAC algorithms with it. It is a portable environment available on a wide range of architectures. Its implementation is built on top of two separate software components: Marcel and Madeleine. Marcel is a POSIX-compliant thread package. Madeleine is a generic communication interface which can be used on top of different communication protocols such as VIA, BIP, SBP, SCI, MPI, PVM and TCP. In our experiments, we used it over TCP.

The second environment is a multi-protocol version of MPICH which also uses Marcel. Thus, it provides multi-threading functionalities inside an MPI implementation. This environment has been chosen according to the wide use of MPI and the fact that it is thread safe. The comparison with PM2 is relevant since although they use the same thread manager, their communications are performed using different schemes (explicit communication with MPI and RPC (Remote Procedure Call) with PM2).

Finally, the last environment is the free Corba [98] ORB OmniOrb 4. It is a robust and high performance ORB which is certified to be in compliance with Corba 2.1. It may be surprising to use an ORB to implement parallel algorithms since this does not correspond to what they are designed for. Nevertheless, Corba in general and OmniOrb 4 in particular both present the minimal features necessary to make those implementations: a communication system between machines and a multi-threaded environment. Our choice has been to take this particular ORB since we already had to use it in another context, so we know it well. Moreover, it has one of the most efficient communication management among the free available ORBs. The comparison with the two other environments is also relevant since it uses a different thread manager and an object oriented communication scheme.

6.1.1 Comparison of the environments

Those environments are multithreaded and, consequently, allow us to implement AIAC algorithms. In order to compare them we have defined three interesting measures: the performances obtained for each of them, the ease of implementing those algorithms and the ease of deploying them onto the parallel systems. We do not present the whole comparison which is quite long but we prefer to highlight the interesting issues. Two typical scientific applications have been tested: a sparse linear system and a nonlinear chemical problem. For more details on those applications, interested readers are invited to consult [16].

6.1.1.1   Performances

Using a low communication network, we have remarked that the three asynchronous implementations (with MPI/Madeleine, Corba and PM2) are always faster than the classical synchronous ones with a standard MPI implementation. Concerning the asynchronous versions, although the environments obtain different results, the ranges of execution times are not that large. So, it can be said that the tested environments globally have the same behavior with AIAC algorithms. We have also concluded that it was impossible to implement exactly the same mechanism from one environment to another. In fact, some of the mechanisms used, such as threads or communications management, are directly dependent on the environments. Unfortunately, it is not always possible to have clear explanations of how those mechanisms are actually implemented. Hence, our algorithms are not exactly implemented in the same way for all the environments because some of the functionalities they require (threads, communications) are not usable in the same way. The main differences between the tested implementations are summarized in Table 6.1.

Sparse linear problem	
PM2	one sending thread receiving threads created on demand
MPI/Mad	one sending thread one receiving thread
OmniOrb 4	N sending threads receiving threads created on demand
Nonlinear problem	
PM2	two sending threads one receiving thread
MPI/Mad	two sending threads two receiving threads
OmniOrb 4	two sending threads receiving threads created on demand

Table 6.1: Differences between the implementations (N is the number of processors).

6.1.1.2   Ease of programming

Obviously, the ease of programming is rather subjective to each user. Here, the comparison is based on the major programming aspects intervening in the programming of AIAC algorithms according to our personal experience.

The first remark is that all the tested environments present both advantages and drawbacks. MPI/Mad is probably the easiest to program since all communications are done in the simple and well known MPI form and the threads are quite easily managed with the Marcel library. PM2 and OmniOrb 4 are quite similar in the way of managing the threads. Concerning the communications, both of them use a remote procedure call mechanism. In PM2, there is a system of explicit data packing before the call to the remote function whereas, in OmniOrb 4, the data to be sent are given as arguments of the called function.

Nevertheless, after having used a dedicated environment for AIAC algorithm programming, all those considerations are meaningless since a dedicated environment provides all the tools to easily develop an AIAC algorithm.

6.1.1.3 Ease of deployment

In terms of deployment, the advantage clearly goes to OmniOrb 4 due to its high flexibility of use over multiple sites and more generally over the global grid that represents the Internet.

Concerning PM2, its deployment is much more restrictive since it requires a complete interconnection graph of the cluster to be used. Moreover, its portability is less important since it does not completely support the use of several systems and/or several architectures of machine in the same cluster.

Concerning the MPI/Mad environment, it is quite similar to PM2 in terms of deployment and portability. Indeed, all the machines must be visible to each other and data representations must be taken into account by the programmer.

Finally, compared to JACE which is portable and easy to deploy, all three environments are not so easily deployable.

After having described the three programming environments that we have initially used to develop AIAC algorithms, we can detail two dedicated environments to efficiently and quickly design AIAC algorithms.

6.2 Two environments dedicated to asynchronous iterative algorithms

The previous environments require the programmer to explicitly manage asynchronous features. Two programming and execution environments have been especially designed for parallel asynchronous iterative algorithms. As those environments are used in the following experiments and as they simplify the development of a parallel iterative algorithm it seems mandatory to us to succinctly present them.

First of all we present JACE. Then, we present Crac which may be viewed as a C++ implementation of JACE with some enhancements. In our experiments, from the programming point of view they are quite similar to use although from the implementation point of view, there are numerous differences which are not important for the designer of AIAC algorithms.

6.2.1   JACE

JACE has been mainly designed and implemented by K. Mazouzi during his PhD thesis. As this environment has been enhanced, we directly present the last version.

JACE [24] is a Java programming and executing environment that permits us to implement efficient asynchronous algorithms as simply as possible. It builds a distributed virtual machine, composed of heterogeneous machines scattered over several distant sites. It proposes a simple programming interface to implement applications using the message passing model. The interface completely hides the mechanisms related to asynchronism, especially the communication manager and the global convergence control. In order to propose a more generic environment, JACE also provides primitives to implement synchronous algorithms and a simple mechanism to swap from one mode to another. JACE relies on three components: *the daemon, the computing task* and *the spawner.*

6.2.1.1   The daemon

The daemon is the entity responsible for executing user applications. It is a Java process running on each node taking part in the computation. Figure 6.1 shows the internal architecture of the daemon which is composed of three layers: *the RMI service, the application layer* and *the communication layer.*

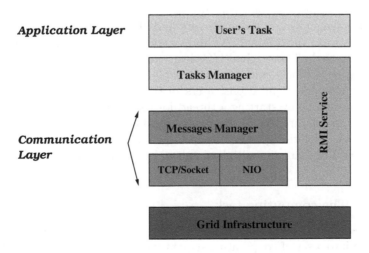

FIGURE 6.1: JACE daemon architecture.

- **RMI Service:** When a daemon is launched, an RMI server is started on it and is continuously waiting for remote invocations. The server provides communications between the daemons and the spawner. It is used to manage the JACE environment as for instance: initializing the daemons, monitoring them or gathering the results.

- **Application Layer:** This layer provides tasks execution and global convergence detection. A daemon may execute multiple tasks, allowing it to reduce distant communications. JACE is designed to control the global convergence process in a transparent way. Tasks only compute their local convergence state and call the JACE API to retrieve the global state. The internal mechanisms of the convergence detection depend on the execution mode, i.e., synchronous or asynchronous.

- **Communication Layer:** Communications between tasks are performed using the message/object passing model. JACE uses waiting queues to store incoming/outgoing messages and two threads (`sender` and `receiver`) to deal with communications. According to the kind of algorithm used, synchronous or asynchronous, the management of queues is different. For a synchronous execution, all messages sent by a task must be received by the other tasks, whereas on an asynchronous execution, only the most recent occurrence of a message, with the same source or destination and containing the same type of information, is kept in the queues. The older one, if it exists, is deleted.

For scalability issues and to achieve better performances, the communication layer should use an efficient protocol to exchange data between remote tasks. For this reason JACE is based on several protocols: TCP/IP Sockets, NIO (New Input/Output) [74, 99] and RMI. NIO is a Java API (introduced in Java 1.4). It provides new features and improved performances in the areas of buffer management, scalable network and file I/O. The most important distinction between the original I/O library and NIO is how the data are packed and transmitted. The original I/O deals with data in streams whereas NIO deals with data blocks and consumes a block of data in one step. Furthermore, for network applications, users previously had to deal with multiple socket connections by starting a thread for each connection. Inevitably, they may have encountered issues such as operating system limits, deadlocks, or thread safety violations, especially in large scale contexts. With NIO, selectors are used to manage multiple simultaneous socket connections on a single thread.

6.2.1.2   The computing task

As in MPI-like environments, the programmer decomposes the problem to be solved into a set of cooperating sequential tasks. These tasks are executed on the available processors and invoke special routines to send or receive messages. A `task` is the computing unit in JACE, which is executed like a thread rather than a process. Thus, multiple tasks may be executed in the same daemon and can share the system resources.

To write a JACE application, the user simply needs to extend the `Task` class and to define a `run()` method containing its program code. The `Task` class may be considered as the programming interface of JACE. It contains a limited set of methods and attributes dedicated to implement asynchronous/synchronous algorithms in a message passing style. To summarize, we can find:

- the nonblocking send/receive,

- the blocking send/receive (for synchronism),

- the global communications: barrier, broadcast,

- the convergence control,

- the finalization.

We also point out here that JACE implementation relies on the Java object serialization to transparently send objects rather than raw data.

6.2.1.3 The spawner

The spawner is the entity that effectively starts the user application. After starting daemons on all nodes, computations begin by launching the spawner program with the following parameters:

- the number of tasks to be executed

- the URL of the task byte-code

- the parameters of the application

- the list of target daemons

- the mapping algorithm (round robin, best effort)

Then, the spawner broadcasts this information to all the daemons. As JACE is designed to execute applications on large scale architectures with a large number of nodes, that is achieved by using an efficient broadcast algorithm based on a binomial tree [67]. This algorithm provides better performances compared to a binary tree.

Let us suppose that we have a binomial tree of 2^d nodes and that node 0 needs to send a message to all other nodes. This algorithm uses d parallel phases where at phase k ($1 \leq k \leq d$) node i (with $i < 2^{d-1}$) sends a message to node $i + 2^{k-1}$. In the general case, the spawning procedure is achieved in $log_2(n)$ communication steps on n nodes. Figure 6.2 presents a binomial tree broadcast procedure with 2^3 nodes.

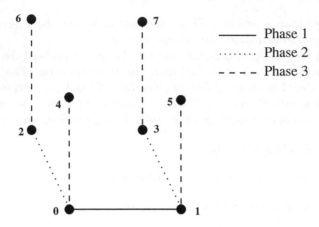

FIGURE 6.2: A binomial tree broadcast procedure with 2^3 elements.

Now, when a task is spawned, an identification number (task ID) is assigned to it. This number is an integer whose value ranges from 0 to $p - 1$, with p being the global number of tasks in the JACE application. This task mapping is done by JACE and by default uses a round robin algorithm. Another method can be used (called best effort) trying to balance the number of tasks over the set of machines. To illustrate these two policies, let us assume 6 tasks $(0, 1 \ldots 5)$ to be mapped on 3 processors. With a round robin algorithm tasks 0 and 3 are mapped on processor 0 and so on. With a best effort algorithm tasks 0 and 1 are mapped on processor 0 and so on. Since communications often take place between consecutive tasks the best effort policy encourages local communications and can be interesting when using multi-processor machines.

6.2.2   CRAC

CRAC has mainly been developed by S. Domas. It has been coded in C++ in order to obtain good performances. It should be noted that for an application coded with JACE in Java and the same application coded with MPI in C (with MPICH/Madeleine), the amount of Java code is one-third smaller than the C code since there is no need to explicitly implement the asynchronism mechanisms. Furthermore, we have obtained an average ratio of 6 between the Java and C execution times, even though it is common to have 10 for scientific applications. However, this ratio is often considered too big in respect to the higher coding effort needed in C. This is why CRAC was developed in a C++ environment, based on the same principles as JACE, and adding some optimized primitives and mechanisms which take into account the architecture of the grid.

6.2.2.1 Architecture

CRAC is based on the classical MPI triplet: daemon, application, spawner. The daemon is launched on each machine constituting the Virtual Distributed Machine (VDM). The user develops its application and launches it with the spawner on the desired machines. However, the similarity with MPI nearly stops here. Even if the CRAC programming interface uses the message passing paradigm, the semantic of communications is completely different and several primitives do not exist in MPI. Furthermore, the internals of CRAC are based on multithreading and even the application is a thread. Finally, the virtual distributed machine relies on a hierarchical view of the network in order to reach machines with private IPs and to limit the bandwidth use on slow links.

The following items present the different components of CRAC, from the VDM to the programming interface.

- Virtual Distributed Machine (VDM): the efficiency problem of distributed executions partly comes from *low* bandwidth on links between distant geographical sites. In that case, a primitive like a gather-scatter that does not take care of the network architecture may be totally inefficient if all messages must take the slowest link in the architecture. Assuming that machines can be gathered in *sites*, which have good bandwidths, and that sites are linked by *low* bandwidths, all global communications may be optimized to take this organization into account. This is the case when the architecture is composed of clusters linked by the Internet. Unfortunately, cluster machines often have private IPs and can only be reached through a frontal machine. To get round that problem, the frontal may relay messages.

MPI does not take into account the network architecture but CRAC does. Thus we can give the following definition of a *site* as a pool of machines which can directly connect to each other. This notion is not necessarily geographical but this may sometimes be the case. For example, if the machines of two distant clusters can freely interconnect, it is better to separate them in two sites if the bandwidth between the clusters is low. If it is similar to the bandwidth inside clusters, they can be gathered in the same site.

Within a site, four types of machines are possible: master, supermaster, slave, frontal. The last type may be applied to any of the first three. For example, a machine may be a slave frontal. This characterization allows us to optimize the management of the VDM (starting/stopping the daemons, spawning, ...), and to reach machines with private IPs.

Here are the definitions of the types:

- frontal: a machine that can relay messages from outside the site to the private IP machines of the site. It can also relay messages to another site if a machine cannot send data outside the site.

- slave: a machine with no particular role (except computations).

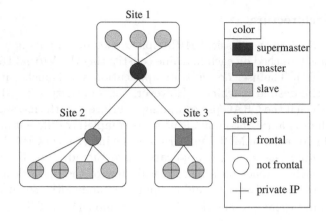

FIGURE 6.3: An example of VDM.

- master: a machine that collects information from the slaves of the site and relays them to the supermaster, or that relays information from the supermaster to the slaves.

- supermaster: a machine that collects/sends information from/to the masters. Obviously, the supermaster is a master but is unique.

The VDM is defined via an XML file, which is a perfect language to describe its hierarchical organization. This file is passed as an argument to a booter (like lamboot) that launches the daemons on each machine of the VDM. Then, a TCP connection is created between each master and the supermaster, and between each slave and its own master. This hierarchy allows us to limit the bandwidth used between the sites. For example, when tasks are spawned for an execution, the supermaster sends the configuration of the execution (the machines used and their number) to all masters, which relay the information to their own slaves.

In order to limit the number of connections between tasks, the convergence detection mechanism also uses this hierarchy. Thus, even if a master runs no task, its daemon is in charge of collecting the local convergence state of each task running in the site.

Figure 6.3 shows an example of VDM with 3 sites. The lines represent the TCP connections that constitute the hierarchical network used for convergence detection and for management (essentially launching and stopping tasks). In sites 2 and 3, two slaves have private IPs. Thus, it is mandatory for a machine of those sites to be a frontal. It may be the master itself as in site 3 or simply another slave, as in site 2. It can be noticed that there are no connections between masters and that the supermaster may also have slaves, as in site 1.

During the execution of an application, a task may communicate data to a task on another machine. The hierarchical network is never used for that.

Instead, a new TCP connection is created between the machines running the two tasks the first time they want to communicate (see below).

- Daemon: a CRAC daemon is launched on each machine of the VDM. During an execution, its main role is to send and receive messages for the local tasks. If the machine is a frontal, the daemon may also relay messages to tasks hosted by another daemon. Those operations are executed by two threads:

- the **Sender** thread: each time it awakes, it checks in the **outgoing queue** the presence of messages to send. If no socket exists to the destination machine, the Sender tries to connect and to retrieve a new socket dedicated to send application data to that destination. Even if the destination machine hosts several tasks, a single socket is used.

  However, the destination machine may have a private IP. In that case, the Sender tries to connect to the frontal machine of the destination site. Each message will be sent to the frontal, which will relay the data to the real destination.

  In order to optimize the global communication time, each message is composed of a header followed by packets and is not sent in one chunk. As each packet has a fixed destination, the Sender performs a loop on the destinations of the packets: it sends a packet to one destination after another. Obviously, if a new message to an existing destination is inserted in the outgoing queue, it must wait for the end of the emission of the current one. But if the new message is for a new destination, it can be sent immediately. That process is a kind of pipeline which greatly reduces the time needed by the last message inserted in the queue to arrive completely at its destination.

- the **Receiver** thread: it uses a polling mechanism to passively detect connection demands and the incoming of data on existing sockets. In the last case, the Receiver uses the header to determine the destination task. If this task is not running on the machine, it means that the message must be relayed and it is directly put in the outgoing queue to be sent by the Sender. If the machine hosts the destination task, the Receiver retrieves the source task from the header and a slot of the **incoming queue**, associated to that task, is used to store the data.

  The slot allocation policy is as follows: The Receiver always checks if a slot with the same message characteristics {source,destination,tag} exists. If this is the case, existing data are overlapped by the ones to come, otherwise a new slot is created. This overlapping is particularly useful to accelerate convergence in asynchronous executions. The slot is freed when the task retrieves its data.

  Taking the example of Figure 6.4, the Receiver of B would create a slot for the first message coming from A. At the end of its first iteration,

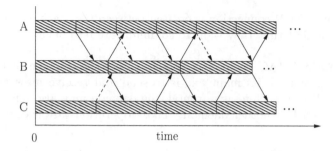

FIGURE 6.4: An example illustrating that some messages (represented with dashed lines) are ignored.

B retrieves the data from the slot. During its second iteration, the Receiver creates a slot for the second message from A but uses the same slot to store the data of the third message. Thus, B is insured to always have the latest data sent by A.

It must be noticed that this policy works perfectly well for synchronous executions. Indeed, for a given triplet {source,destination,tag}, a single message can be sent/received during the same iteration. Thus, there cannot be lost data because of overlapping messages.

The daemon also creates the **Converger** thread that is in charge of collecting and updating information about the convergence, using the hierarchical network of the VDM. It implies that the supermaster has more information to collect than the masters, and the masters more than the slaves. Thus, the work of this thread depends on the machine type but whatever the case, its final goal is to provide the global convergence state to the tasks.

- Task: the application task is a thread that is executed within the daemon context. Thus, the task can directly access message queues (incoming and outgoing). This is not the case for MPI, in which a task is a process and must communicate (with a Unix socket or shared memory) with the daemon to send/receive data.

As CRAC is an object environment, the **Task** class is defined as a thread, containing all primitives of the programming interface and the classical attributes of a task (identifier, number of tasks in the daemon and in the VDM,...). CRAC also declares (as an include file) the **UserTask** class which inherits from **Task**. This class contains a run() method that must be defined by the user in a C++ file, which is compiled as a shared library. When a task is launched on a machine via the spawner, the daemon dynamically loads its code and creates a new thread object containing that code. The thread is started and its run() method automatically called, as in Java.

- **Spawner:** the CRAC spawner is a classical MPI spawner, except that it uses an XML file to specify which and how many tasks are launched on which machine. The access path of the code of each task must be given for each machine. Thus, it is possible to have an MIMD execution. It is also possible to pass arguments to each task. For now, the spawning is only static and tasks cannot be added during an execution.

- **Programming interface:** it is defined in the Task class. It provides the classical primitives to implement message passing codes but some have special semantics and some are dedicated to iterative algorithms. Here are four characteristic examples that greatly differ from MPI:

- CRACSend(): the emission of a message is never blocking. This routine simply copies the data in a slot of the outgoing queue. Thus, the buffer containing the data can be immediately reused. The slot allocation policy is identical to that of the incoming queue: a new slot may be created or an existing slot chosen and its data overlapped.

- CRACRecv(): the reception may be blocking or not, depending on a parameter of this function. In MPI, the nonblocking reception returns an identifier that allows the receiver to test and to wait for the total reception of the message. In CRAC, it is like a test/receive. If the message is in the incoming queue, the buffer passed to CRACRecv() is filled and it is left empty if no message arrived. This semantic is dedicated to an asynchronous execution for which it must be possible to begin another iteration without new data being received.

- CRACConvergence(): it may be blocking or not, depending on a parameter of that function. In both cases, it takes a boolean as a parameter, which is the local convergence state. It returns the global convergence state as a boolean. Obviously, that routine must be used in blocking mode for a synchronous execution. For a description of its work in asynchronous mode, one can refer to [22].

- tags: each message must be marked by an integer value defined by the user. As mentioned above and in the Receiver description, there is an automatic replacement if a message with the same triplet {source,destination,tag} is already present in the queues. Thus, the user must assign the same tag to the messages that are used to update the same data set of the destination task. Obviously, the same tag must never be used for messages updating different data sets.

6.3   Ratio between computation time and communication time

Scientists who work with parallel architectures define the ratio between the computation time and the communication time as the granularity of an application. This comes from the fact that in traditional algorithms (i.e., synchronous algorithms, iterative or not), periods of computation are typically separated from periods of communication by synchronizations. However, dealing with AIAC algorithms, there are no more synchronizations, thus that traditional definition is not very appropriate. In order to be scalable, an algorithm must be coarse grained, i.e., have long computation parts without communications. That is to say, messages must be gathered in order to reduce the number of communications. It is straightforward to see that algorithms based on the multisplitting method are coarse grained. Nevertheless, even though an algorithm has been designed to be coarse-grained, the ratio between the computation time and the communication time may affect its execution times in a given computing context. Thus, in our opinion, this ratio is more important than the notion of the granularity of an application. The main reason lies in the fact that the granularity definition does not involve the speed of the interconnection network which affects the communication time. In the following, as we will see, for some algorithms in some computing contexts the synchronous version will be faster than the asynchronous one. Of course, we shall also see the opposite situation.

6.4   Experiments in the context of linear systems

As explained in the respective chapters on synchronous and asynchronous iterative algorithms, the multisplitting method is very efficient to solve linear systems, especially sparse ones. That kind of systems arises in numerous modelizations.

6.4.1   Context of experimentation

In this subsection we explain all the experiments we have performed and we analyze the results obtained. Experiments have been conducted on the GRID'5000 architecture, a nation-wide experimental grid in France [35]. Currently, the GRID'5000 platform is composed of an average of 1300 bi-processors which are located in 9 sites in France: Bordeaux, Grenoble, Lille, Lyon, Nancy, Orsay, Rennes, Sophia-Antipolis, Toulouse, as described in Figure 6.5. Most of those sites have a Gigabit Ethernet Network between the local machines.

The links between the different sites range from 2.5 Gbps up to 10 Gbps. Most of the processors in the platform are AMD Opterons. For more details on the GRID'5000 architecture, interested readers are invited to visit the website: http://www.grid5000.fr. The architecture of that platform constantly evolves as each site independently updates its local architecture.

Concerning the processors used in the following experiments, they range from AMD Opterons 246 2 GHz to AMD Opterons 250 2.4 GHz.

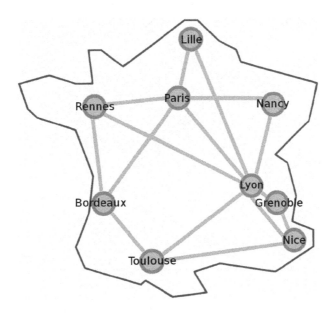

FIGURE 6.5: The GRID'5000 platform in France.

In order to compare the behavior of both synchronous and asynchronous multisplitting methods for linear systems, we have built a square matrix generator which ensures that the corresponding linear problem can be solved by those methods, i.e., their convergence is guaranteed by the matrix properties.

The matrices we have used are built using the following scheme. The diagonal of the matrix is not empty nor are its two neighbor diagonals. Then, according to the number of diagonals specified by the user, some of the other diagonals are not empty. Those diagonals are equitably scattered between the diagonal of the matrix and a bandwidth specified by the user. Consequently, in the experiments we report those two parameters (number of diagonals and bandwidth). Off-diagonal nonempty elements of a matrix are negative random values with a value between -1 and 0. Diagonal elements are equal to the inverse of the sum of the nonempty elements of the same line plus a random value whose interval is defined by the user. This allows us to change the spectral radius of the iteration matrix which acts on the number of iterations required to reach a given threshold during the resolution. Such generated matrices are M-matrices for which it is known that multisplitting algorithms converge. Figure 6.6 illustrates the case where the bandwidth is equal to half the matrix size with 7 nonempty diagonals.

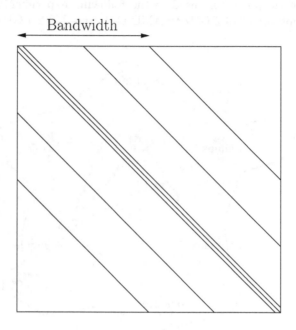

FIGURE 6.6: Example of a generated square matrix.

In the following experiments we mention the size of the square matrix, its parameters, the sequential solver used, the execution mode (synchronous or asynchronous), the execution time in seconds, the number of iterations and the number of processors used with their location and characteristics. It should be noticed that the number of iterations in the synchronous mode is always constant whereas it is not the case in the asynchronous mode since it varies from one execution to another and from one processor to another, mainly according to the possible network fluctuations and to the relative powers of the processors if they are heterogeneous. That is why, in this case, we report an interval with the minimum and the maximum number of iterations according to the execution and the processors. Every reported result is the mean value of a series of four executions. Finally, it should be noticed that we have not experimented with matrices for which the resulting iteration matrices have too small a spectral radius because in such cases, the number of iterations would be smaller and not very representative. Therefore we have chosen quite complex matrices.

In our experiments, we did not change the default parameters of the MUMPS [3, 4] and SuperLU [82, 70] packages. With SparseLib we have chosen the GMRES method with a ILU preconditioner. In order to implement those synchronous and asynchronous algorithms we have used the CRAC library [40].

6.4.2 Comparison of local and distant executions

In this first series of experiments, we wanted to compare the execution times of the multisplitting method using only local nodes, and using the same number of nodes scattered over distant sites. To achieve this, we have used a square matrix of size 10,000,000 with 70 processors located either in Nice for the local case or 70 processors scattered as follows: 30 in Orsay, 20 in Lille and 20 in Nice for the distant one. Table 6.2 shows the execution times obtained with the local cluster in Nice. In this table we can remark that with large bandwidth matrices, the multisplitting method is more efficient in the asynchronous mode than in the synchronous one. This can be explained by the fact that the larger the bandwidth is, the more communications are required with the more neighbors. With a smaller bandwidth, the synchronous version is faster. It can also be noticed that the number of iterations required to reach the convergence in the asynchronous mode is always greater than for the synchronous mode. This remark will always be true in the following.

Solver	Synchronous		Asynchronous	
	exec. time (s)	nb. iter.	exec. time (s)	nb. iter.
13 diagonals, bandwidth: 5,000,000				
SparseLib	88.69	142	57.42	[207-296]
MUMPS	98.73	142	70.39	[198-280]
SuperLU	80.23	142	49.00	[241-365]
13 diagonals, bandwidth: 1,000,000				
SparseLib	79.89	125	57.60	[182-247]
MUMPS	98.33	125	69.75	[174-237]
SuperLU	72.87	125	50.92	[183-255]
13 diagonals, bandwidth: 100,000				
SparseLib	39.19	51	48.01	[57-75]
MUMPS	15.45	51	19.81	[65-106]
SuperLU	12.40	51	15.21	[71-111]

Table 6.2: Execution times of the multisplitting method coupled to different sequential solvers for a generated square matrix of size 10.10^6 with 70 machines in a local cluster (Sophia).

Table 6.3 shows the execution times of the same matrices as in Table 6.2 with the same number of nodes but located in three sites. The execution times are higher than in a local context. This is not surprising since distant communications take more time than local ones. With large bandwidth matrices execution times are much longer and the comparison between a local running and a distant one may seem irrelevant. In this case, the use of a distant cluster is limited to solving large square matrices that cannot be solved using a local cluster.

Nevertheless, the asynchronous version is more robust to distant communications than the synchronous one. This is due to the implicit overlapping of communications by computations inherent to the asynchronous model. However, with smaller bandwidth matrices the behavior of the solver between a local and a distant running is more comparable. The ratio between the local and the distant running is bounded by three although the smallest machines in the distant configuration are the same as in the local one. Another issue is that whatever the sequential solver used in our experiments is, the execution times are relatively similar for this size of matrix.

Solver	Synchronous		Asynchronous	
	exec. time (s)	nb. iter.	exec. time (s)	nb. iter.
13 diagonals, bandwidth: 5,000,000				
SparseLib	1,340.12	142	770.62	[1,821-2,354]
MUMPS	1,178.56	142	741.65	[1,582-2,101]
SuperLU	1,109.12	142	736.63	[1,782-2,095]
13 diagonals, bandwidth: 1,000,000				
SparseLib	1,244.25	125	517.69	[1,876-2,320]
MUMPS	1,318.63	125	512.32	[2,019-2,764]
SuperLU	1,298.71	125	506.76	[2,102-2,908]
13 diagonals, bandwidth: 100,000				
SparseLib	83.97	51	48.73	[65-86]
MUMPS	60.48	51	42.29	[178-279]
SuperLU	62.35	51	46.56	[283-422]

Table 6.3: Execution times of the multisplitting method coupled to different sequential solvers for a generated square matrix of size 10.10^6 with 70 machines located in 3 sites (30 in Orsay, 20 in Lille and 20 in Sophia).

To sum up this first series of experiments, the ratio between the computation time and the communication time plays a major role. With those matrices, when the bandwidth increases, the computation time stays quite of the same order whereas the communication time drastically increases. That is why, with large bandwidth matrices, the multisplitting method is less efficient in a distant context. But it should be noticed that any other solvers would have the same caveat, probably with a stronger effect due to the amount of communications and synchronizations. From a general point of view, the higher this ratio between the computation time and the communication time is, the more favored the synchronous version is compared to the asynchronous one, and reciprocally. That is why when executing an algorithm in a local cluster, the synchronous version may be faster than the asynchronous one. Conversely, executing the same algorithm within a grid context, where communication performances are worse, the communication time would be longer and the ratio would often decrease in favor of the asynchronism.

6.4.3 Impact of the computation amount

In the previous series of experiments we have shown that the communication time has a great influence on the execution time especially in a distant context. In this second series of experiments we want to study how the execution time is affected when the computation amount increases. In Table 6.4, we report the experiments with different sizes of square matrices. For each one, the bandwidth is quite small, which ensures that the execution times will not be too long. For those experiments we have used 120 machines scattered over 4 sites (40 in Rennes, 40 in Paris, 25 in Nancy and 15 in Lille). As our goal was not to compare the performances of the different sequential solvers, only one of them has been used in those experiments, the MUMPS solver.

Size of the matrix	Number of diagonals	Bandwidth	Synchronous exec. time (s)	nb. iter.	Asynchronous exec. time (s)	nb. iter.
1.10^6	23	1,000	10.05	72	4.55	[303-475]
2.10^6	23	2,000	14.98	69	5.39	[195-229]
4.10^6	23	4,000	19.33	68	12.31	[204-268]
6.10^6	23	6,000	24.19	69	13.34	[146-176]
8.10^6	23	8,000	27.87	68	18.18	[142-143]
10.10^6	23	5,000	28.10	67	22.22	[136-144]

Table 6.4: Execution times of the multisplitting method coupled to the MUMPS solver for different sizes of generated matrices with 120 machines located in 4 sites (40 in Rennes, 40 in Orsay, 25 in Nancy and 15 in Lille).

Those experiments emphasize that the smaller the size of the square matrix is, the more efficient the asynchronous version is, compared to the synchronous one. This is easily understandable since the computation amount increases with the size of the square matrix. So, the ratio between the computation time and the communication time decreases and the communications become less penalizing. The number of iterations to reach the convergence in the asynchronous version clearly shows that point as that number is larger with matrices of small sizes. When the computation amount becomes more important, the difference between the synchronous and the asynchronous versions decreases. It should be noticed that the network bandwidth between the different sites of the GRID'5000 architecture is very important compared to the traditional network bandwidths. So, with the GRID'5000 platform the ratio between the computation time and the communication time for which algorithms are efficient is very different from more traditional grid environments with low bandwidth networks.

6.4.4 Larger experiments

In a third series of experiments we have used a larger number of processors in order to measure the scalability of the multisplitting method.

Solver	Size of the matrix	Nb. of diag.	Band-width	Synchronous		Asynchronous	
				exec. time (s)	nb. iter.	exec. time (s)	nb. iter.
MUMPS	10.10^6	13	5,000	16.41	30	10.98	[138-159]
SuperLU	10.10^6	13	5,000	19.83	30	14.77	[85-109]
MUMPS	20.10^6	13	5,000	15.91	22	15.48	[74-79]

Table 6.5: Execution times of the multisplitting method coupled to the MUMPS or SuperLU solvers for different sizes of generated matrices with 190 machines located in 5 sites (30 in Rennes, 30 in Sophia, 70 in Orsay, 30 in Lyon and 30 in Lille).

In Table 6.5 we report three configurations for which we have compared the execution times of the two execution modes of the multisplitting method with 190 machines scattered over 5 sites (30 in Rennes, 30 in Sophia, 70 in Orsay, 30 in Lyon and 30 in Lille). It results from those experiments that the two sequential direct solvers which can be used have quite a similar behavior. With the larger matrix size, equal to 20,000,000, the synchronous and the asynchronous versions require about the same amount of time. This is due to the ratio between the computation time and the communication time which does not favor any of the two versions.

In all our previous experiments, we have only studied executions with one computation thread per machine. Let us remind the interested reader that machines in the GRID'5000 architecture are at least bi-processors. As previously mentioned, CRAC is a multithreaded library and it is possible, with the multisplitting method, to run more than one computation task per machine. In fact, as soon as the computation task is thread safe, i.e., it supports being executed by multiple threads, the multisplitting method can be executed with multiple computation tasks. Unfortunately, two of the three internal solvers used are not thread safe (MUMPS and SuperLU); that is why we have only experimented it with SparseLib.

In that experiment, we have used two sites (Paris and Nice) and chosen matrices with larger bandwidths, up to 3,000,000. Table 6.6 shows the results. Although only two distant sites are used, we can remark that the asynchronous version is still faster than the synchronous one. The multisplitting method simply allows us to use multi-processor machines as soon as the sequential linear solver used is thread safe. Hence, it is easy to increase the computing power without any additional effort.

Number of diagonals	Bandwidth	Synchronous		Asynchronous	
		exec. time (s)	nb. iter.	exec. time (s)	nb. iter.
13	300,000	132.51	134	87.14	[634-859]
23	300,000	163.88	141	104.37	[576-809]
13	3,000,000	353.80	142	245.68	[980-1279]

Table 6.6: Execution times of the multisplitting method coupled to the SparseLib solver for generated square matrices of size 30.10^6 with 200 bi-processors located in 2 sites (120 in Paris, 80 in Nice), so 400 CPUs.

6.4.5   Other experiments in the context of linear systems

Using the multisplitting method to solve linear systems has other interesting advantages which are worth noticing. In this subsection, we report some experiments performed to measure some interesting features of the multisplitting method.

6.4.5.1   Influence of the overlapping

In several parts of this book we have mentioned that the overlapping of components may reduce the number of iterations, and consequently, the execution time. In the following experiments we have measured the impact of the overlapping on the number of iterations required to reach the convergence and the impact on the factorization time. We only focus on the multisplitting method with a direct inner solver because the size of each submatrix, which varies depending on the overlapping size, has an important influence on the factorization time. Moreover, we only consider the synchronous version in order to measure the influence on the number of iterations. Considering an asynchronous execution would lead to different numbers of iterations from one execution to another. Finally, we have generated a smaller square matrix of size 100,000 and used a smaller cluster composed of only 10 machines scattered on 2 distant sites in France, connected by a 20 Mbps optic fiber link. The machine configuration ranges from Intel Pentiums IV 1.7 GHz to Intel Pentium IV 2.6 GHz with 512 MB memory.

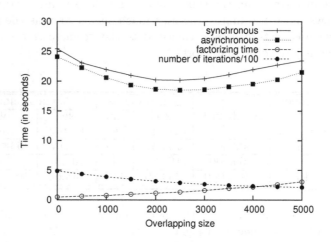

FIGURE 6.7: Impacts of the overlapping for a generated square matrix of size 100000.

Figure 6.7 illustrates the impacts of the overlapping on the proposed algorithms. The larger the overlapping is, the more time-consuming the factorization step is. Consequently, its size should be defined carefully and should take into account the size of the matrix and its parameters such as the factorization time, the spectral radius, etc. In the figure, the numbers of iterations for the synchronous algorithm are reported divided by 100 to simplify the reading of the display. It can be seen that, in that example, the best overlapping size is around 2500.

6.4.5.2   Memory requirements with a direct method

It is also interesting to study the influence of the number of processors on the memory usage when a direct inner solver is used in the multisplitting method to solve linear systems. The factorization step requires a lot of memory; that is why it is interesting to measure the influence of the number of processors on the memory usage. As mentioned before, by splitting the matrix, each processor is in charge of a small part of the matrix. This part is all the smaller as the number of processors is important. Table 6.7 shows the maximal memory usage for a processor in function of the number of processors used. In those experiments on the memory requirements, we have used the synchronous version of the multisplitting method with the SuperLU solver on a cluster composed of 20 Pentiums IV 3 GHz with 1 GB memory. The network is a standard Giga Ethernet with 1 Gbps. We have used a matrix called *cage*12 which can be found in the University of Florida Sparse Matrix Collection [42].

number of processors	execution time (s)	factorization time (s)	maximal memory usage (MB)	number of iterations
6	28570.82	3114.54	1748.81	16
7	11795.22	1866.55	1267.43	16
8	2136.03	1297.61	1061.15	17
9	902.98	886.27	864.40	17
10	420.49	411.17	596.48	17
11	431.71	423.32	576.20	17
12	218.55	212.02	429.86	17
13	186.23	178.97	371.26	17
14	220.47	212.59	363.97	18
15	102.25	98.44	258.59	17
16	133.25	121.03	265.66	18
17	85.62	74.20	214.39	17
18	77.37	71.26	203.34	18
19	47.30	44.47	157.35	18
20	74.33	68.93	176.57	18

Table 6.7: Impacts of memory requirements of the synchronous multisplitting method with SuperLU for the *cage*12 matrix.

This table contains several important points that should be commented upon. No experimentation has been conducted with fewer than 6 processors because the machines did not have enough memory. In those experiments, we only present splittings with equal sizes. According to the dependencies of the matrix and the values of submatrices, the memory required for each processor differs. In the table, we only report the maximum of the memory used because it is significant information. In fact, with that cluster configuration, in which each processor has 1 Gb memory, the memory requirements in the experiments with 6 to 8 processors exceed the total available memory on at least one of the involved machines, which has to swap a part of its memory during the computation. This is why the execution times are so important with those numbers of processors. Besides, we can remark that, contrary to the other processor configurations, i.e., more than 8, there is an important difference between the execution times and the factorization times, which is also due to the memory swapping.

Other experiments with linear systems in different execution contexts may be found in [13, 18, 15, 14].

6.5 Experiments in the context of partial differential equations using a finite difference scheme

In this section, we are interested in solving a nonlinear system of equations which simulates the evolution of the concentrations of two chemical species in a two dimensional domain. This problem corresponds to an advection-diffusion system with two species. It is solved by using a discretization of the space on a two-dimensional grid (x, z).

The evolution of the species concentrations is given by

$$\frac{\partial c^i}{\partial t} = K_h \frac{\partial^2 c^i}{\partial x^2} + V \frac{\partial c^i}{\partial x} + \frac{\partial}{\partial z} K_v(z) \frac{\partial c^i}{\partial z} + R^i(c^1, c^2, t) \tag{6.1}$$

where c^i ($i = 1, 2$) denotes the concentrations of the chemical species, K_h, V and K_v, respectively, denote the horizontal diffusion coefficient, the velocity and the vertical diffusion coefficient. The function $R^i()$ represents the reaction of the chemical species [73]:

$$\begin{aligned} R^1(c^1, c^2, t) &= -q_1 c^1 c^3 - q_2 c^1 c^2 + 2q_3(t)c^3 + q_4(t)c^2 \\ R^2(c^1, c^2, t) &= q_1 c^1 c^3 - q_2 c^1 c^2 + q_4(t)c^2 \end{aligned} \tag{6.2}$$

with

$$\begin{array}{ll} K_h = 4.0 \times 10^{-6} & V = 10^{-3} \\ K_v(z) = 10^{-8} e^{\frac{z}{5}} & c^3 = 3.7 \times 10^{16} \\ q_1 = 1.63 \times 10^{-16} & q_2 = 4.66 \times 10^{-16} \\ q_j(t) = e^{-a_j/sin(\omega t)} & \text{for } sin(\omega t) > 0 \\ q_j(t) = 0 & \text{otherwise} \end{array} \tag{6.3}$$

and $j = 3, 4$, $\omega = \pi/43200$, $a_3 = 22.62$ and $a_4 = 7.601$.

The initial conditions are the following ones:

$$\begin{aligned} c^1(x, z, 0) &= 10^6 \alpha(x)\beta(z) \\ c^2(x, z, 0) &= 10^{12} \alpha(x)\beta(z) \end{aligned} \tag{6.4}$$

with

$$\begin{aligned} \alpha(x) &= 1 - (0.1x - 1)^2 + (0.1x - 1)^4/2 \\ \beta(z) &= 1 - (0.1z - 1)^2 + (0.1z - 4)^4/2 \end{aligned} \tag{6.5}$$

The discretization in space along x and z allows us to rewrite the system of PDEs in Equation (6.1) in a system of ODEs (*Ordinary Differential Equations*) of the form

$$\frac{dy(t)}{dt} = f(y(t), t) \quad \text{with} \quad y = (c^1, c^2) \tag{6.6}$$

where $y(t)$ is a vector of size $2n$ and n is the total number of elements in the discretization grid.

Using the implicit Euler scheme, the previous equation can be rewritten as

$$\frac{y(t+h) - y(t)}{h} = f(y(t+h), t+h) \tag{6.7}$$

where h is the discretization time step.

Solving the previous equation is equivalent to finding $F(y(t+h), y(t), t) = 0$ with

$$F(y(t+h), y(t), t) = y(t) + h * f(y(t+h), t+h) - y(t+h) \tag{6.8}$$

where t and $y(t)$ are known and $y(t+h)$ is the unknown.

In a sequential execution, the Newton algorithm allows us to compute the solution of that equation. So, as soon as the problem has been solved for one instant t, it can then be solved at the next instant $t+h$ according to the given discretization time step h. And that process is repeated until the solution has been computed over the entire considered time interval.

In parallel, using an asynchronous algorithm, it is possible to design two versions. One is based on the asynchronous Newton-multisplitting algorithm presented in Section 5.6.5 and the other one is based on the asynchronous multisplitting-Newton algorithm presented in Section 5.6.6.

As previously seen, the multisplitting-Newton version presents the advantage of not requiring any synchronization. As this problem is nonstationary and involves the resolution of a nonlinear equation at each time step, the use of the multisplitting-Newton method allows one to obtain an algorithm which only requires one synchronization at each discretization time step. Hence, this algorithm may seem suited to a grid computing context with distant clusters.

In the following experiments we have also used the GRID'5000 architecture presented in Section 6.4.1.

In Tables 6.8 and 6.9, we report the results of the multisplitting-Newton method applied to the advection-diffusion problem discretized as described above. In order to sequentially solve the underlying sparse linear systems, we have used the MUMPS software [2] which is a direct sparse linear solver. In those experiments, 120 machines have been used, scattered over 4 sites of the GRID'5000 platform.

In those tables, each result is the mean value of a series of 10 executions. In order to compare the behavior of the application, we have chosen two discretization steps: 360 and 720 seconds. For each value, reported execution times have been achieved for 2 time steps. Different sizes of problems have

Parallel Iterative Algorithms

discretization time step: 360 s				
problem size	synchronous		asynchronous	
	exec. time (s)	nb. of iter.	exec. time (s)	nb. of iter.
1400 × 1000	47.8	252	25.9	[264-290]
2100 × 1500	123.8	429	80.6	[452-496]
2800 × 2000	271.7	626	190.7	[710-832]
4200 × 3000	981.3	984	668.8	[1108-1274]

Table 6.8: Execution times of the multisplitting-Newton method coupled to the MUMPS solver for different sizes of the advection-diffusion problem with 120 machines located in 4 sites and a discretization time step of 360 s.

discretization time step: 720 s				
problem size	synchronous		asynchronous	
	exec. time (s)	nb. of iter.	exec. time (s)	nb. of iter.
1400 × 1000	75.5	393	39.4	[401-437]
2100 × 1500	242.1	696	184.8	[712-846]
2800 × 2000	431.9	964	299.0	[1042-1169]
4200 × 3000	1368.9	1523	1046.7	[1691-1864]

Table 6.9: Execution times of the multisplitting-Newton method coupled to the MUMPS solver for different sizes of the advection-diffusion problem with 120 machines located in 4 sites and a discretization time step of 720 s.

been examined in order to analyze the behavior of CRAC with a variable ratio between computation and communication times. For example, a problem size of 4200 × 3000 means that, because of the two chemical species, the global matrix has $2 \times 4200 \times 3000 = 25,200,000$ rows and columns, with 10 non-null elements per row. As previously mentioned, multisplitting methods allow the overlapping of some components, which may decrease the number of iterations. In our experiments, an overlapping size equal to 20 for each dimension has been chosen.

The study of Tables 6.8 and 6.9 reveals that the asynchronous version of the algorithm is always faster than the synchronous one. This phenomenon is due to the fact that in the synchronous case, all tasks are synchronized at each iteration of the multisplitting method. When the problem size increases, the ratio of the computation time over the communication time also increases, and the difference between the synchronous and the asynchronous execution times decreases. This is clearly shown in the last column of Table 6.10 which gives the synchronous execution time divided by the asynchronous one. This

discretization time step	problem size	exec. times ratio
360	1400 × 1000	1.85
	2100 × 1500	1.53
	2800 × 2000	1.42
	4200 × 3000	1.47
720	1400 × 1000	1.92
	2100 × 1500	1.31
	2800 × 2000	1.44
	4200 × 3000	1.31

Table 6.10: Ratios between synchronous and asynchronous execution times of the multisplitting-Newton method for different sizes and discretization time steps of the advection-diffusion problem with 120 machines located in 4 sites.

fact was commonly observed in all our studies of asynchronous algorithms. For each version of the algorithm, Tables 6.8 and 6.9 also report the number of iterations required to reach the convergence. As already pointed out, in the asynchronous case, that number varies from one execution to another and from one processor to another. That is why, as in the experiments in the context of linear problems, we report an interval which corresponds to the minimum and maximum numbers of iterations for the different executions. Without considering the mode of execution of the algorithm, the larger the size of the discretization step is, the larger the number of iterations required to reach the convergence is.

Other experiments with nonlinear systems in different execution contexts may be found in [21, 20].

Appendix

A-1 Diagonal dominance. Irreducible matrices

DEFINITION A.1 *An $n \times n$ matrix $A = (A_{i,j})_{1 \leq i,j \leq n}$ is diagonally dominant if for all $i \in \{1, ..., n\}$*

$$|A_{i,i}| \geq \sum_{j \neq i} |A_{i,j}|. \tag{A.9}$$

The matrix A is strictly diagonally dominant if strict inequality is valid for all i in (A.9).

DEFINITION A.2 *An $n \times n$ matrix $A = (A_{i,j})_{1 \leq i,j \leq n}$ is reducible if there exists an $n \times n$ permutation matrix P such that*

$$PAP^T = \begin{pmatrix} B & C \\ 0 & D \end{pmatrix},$$

where B is an $r \times r$ submatrix and C is a $n - r \times n - r$ submatrix. If no such permutation matrix exists, then A is said to be irreducible.

DEFINITION A.3 *An $n \times n$ matrix $A = (A_{i,j})_{1 \leq i,j \leq n}$ is irreducibly diagonally dominant if A is irreducible, diagonally dominant and strict inequality holds in (A.9) for at least one i.*

THEOREM A.1
Let A be an $n \times n$ real or complex matrix. If A is either strictly or irreducibly diagonally dominant then it is invertible.

PROOF   See, e.g., [93]   □

DEFINITION A.4 *A real $n \times n$ matrix $A = (A_{i,j})_{1 \leq i,j \leq n}$ is positive semidefinite if*

$$x^T A x \geq 0, \ \forall x \in \mathbb{R}^n.$$

It is positive definite if strict inequality holds whenever $x \neq 0$.

The eigenvalues of a symmetric positive (semi)definite matrix are positive (nonnegative).

PROPOSITION A.1
Let A be a real $n \times n$ matrix. If A is symmetric, irreducibly diagonally dominant and has positive diagonal elements then A is positive definite.

A-1.1　Z-matrices, M-matrices and H-matrices

DEFINITION A.5　*An $n \times n$ square real matrix A is a Z-matrix if for any $i, j \in \{1, ..., n\}$, $A_{i,i} > 0$ and $A_{i,j} \leq 0$ for $i \neq j$.*

PROPOSITION A.2
Let A be a Z-matrix, then the following properties are equivalent:

1. *There exists a nonnegative vector u $(u \geq 0)$ such that $Au > 0$.*

2. *There exists a positive vector u $(u > 0)$ such that $Au > 0$.*

3. *The matrix A is nonsingular and $A^{-1} \geq 0$.*

4. *The spectral radius of the Jacobi matrix associated to A is strictly less than 1, i.e.,*
$$\rho(I - D^{-1}A) < 1,$$
where D is the diagonal part of A.

PROOF　See Fieder et al. [53].　　　　　　　　　　　　　　　　$\Box$

DEFINITION A.6　*An M-matrix is a Z-matrix which satisfies the properties of Proposition A.2.*

It can be deduced that an M-matrix A satisfies the maximum principle,
$$Au \leq 0 \Rightarrow u \leq 0.$$

Let us associate to the real matrix A the comparison matrix $\langle A \rangle$ the coefficients $\bar{a}_{i,j}$ of which satisfy
$$\bar{a}_{i,i} = a_{i,i} \ and \ \bar{a}_{i,j} = -|a_{i,j}| \ if \ i \neq j$$

DEFINITION A.7　*A will be called an H-matrix if the matrix $\langle A \rangle$ is an M-matrix.*

We can see that M-matrices are special cases of H-matrices.

LEMMA A.1

Let B be a real $n \times n$ matrix and assume that $\rho(B) < 1$. Then $(I - B)^{-1}$ exists and

$$(I - B)^{-1} = \lim_{k \to \infty} \sum_{i=0}^{k} B^i.$$

A-1.2 Perron-Frobenius theorem

THEOREM A.2

Let $A \geq 0$ be an irreducible $n \times n$ matrix. Then,

1. *A has a positive real eigenvalue equal to its spectral radius.*

2. *To $\rho(A)$ there corresponds an eigenvector $x > 0$.*

3. *$\rho(A)$ increases when any entry of A increases.*

4. *$\rho(A)$ is a simple eigenvalue of A.*

PROOF [97], [58]. $\quad\quad\quad\quad\quad\quad\quad\quad\quad\quad\quad\quad\quad\quad\quad\quad\quad$ ⬜

A-1.3 Sequences and sets

A sequence $\{x^{(k)}\}_{k \in \mathbb{N}}$ of complex numbers is said to converge to a number x^* if for arbitrary small and positive number ε, there exists an integer I such that for any $k \geq I$ we have $|x^{(k)} - x^*| < \varepsilon$.

A real sequence $\{x^{(k)}\}_{k \in \mathbb{N}}$ is said to converge to $+\infty$ (respectively $-\infty$) if for every $M \in \mathbb{R}$, there exists I such that $x^{(k)} \geq M$ (respectively $x^{(k)} \leq M$) for $k \geq I$.

A real sequence $\{x^{(k)}\}_{k \in \mathbb{N}}$ is called *bounded above* (respectively *below*) if there exists some real M such that $x^{(k)} \leq M$ (respectively $x^{(k)} \geq M$) for all k.

A real sequence $\{x^{(k)}\}_{k \in \mathbb{N}}$ is *bounded* if the sequence $\{|x^{(k)}|\}_{k \in \mathbb{N}}$ is bounded above.

A real sequence $\{x^{(k)}\}_{k \in \mathbb{N}}$ is said to be *nonincreasing* (respectively *nondecreasing*) if $x^{(k+1)} \leq x^{(k)}$ (respectively $x^{(k+1)} \geq x^{(k)}$) for all k.

PROPOSITION A.3

Every bounded nonincreasing or nondecreasing real sequence converges to a finite real number.

Given a sequence $\left\{x^{(k)}\right\}_{k\in\mathbb{N}}$, the *supremum* and the *infimum* of $\left\{x^{(k)}\right\}_{k\in\mathbb{N}}$ are defined by

$$\sup_k x^{(k)} = \sup\left\{x^{(k)},\ k\in\mathbb{N}\right\} \text{ and } \inf_k x^{(k)} = \inf\left\{x^{(k)},\ k\in\mathbb{N}\right\}.$$

Define $y^{(m)} = \sup\left\{x^{(k)},\ k\geq m\right\}$ and $z^{(m)} = \inf\left\{x^{(k)},\ k\geq m\right\}$, then the sequences $\left\{y^{(m)}\right\}_{m\in\mathbb{N}}$ and $\left\{z^{(m)}\right\}_{m\in\mathbb{N}}$ are, respectively, nonincreasing and nondecreasing so they converge to possibly infinite numbers. We have the following result.

PROPOSITION A.4

Let $\left\{x^{(k)}\right\}_{k\in\mathbb{N}}$ *be a real sequence, then*

1. $\inf_k x^{(k)} \leq \lim_{k\to\infty} \inf_k x^{(k)} \leq \lim_{k\to\infty} \sup_k x^{(k)} \leq \sup_k x^{(k)}.$

2. $\left\{x^{(k)}\right\}_{k\in\mathbb{N}}$ *converges to* x^* *if*
   $\lim_{k\to\infty} \inf_k x^{(k)} = \lim_{k\to\infty} \sup_k x^{(k)} = x^*.$

A vectorial sequence $\left\{x^{(k)}\right\}_{k\in\mathbb{N}}$, $x^{(k)} \in \mathbb{C}^n$ is said to converge to $x^* \in \mathbb{C}^n$ if the *ith* coordinate of $x^{(k)}$ converges to the *ith* coordinate of x^*.

DEFINITION A.8   *Let* B *be a subset of* $\mathbb{C}^n$. *We say that a vector* $x \in \mathbb{C}^n$ *is a limit point of* B *if it is a limit of a sequence* $\left\{x^{(k)}\right\}_{k\in\mathbb{N}}$ *of elements of* B.

DEFINITION A.9   *A set* $B \subset \mathbb{C}^n$ *is called closed if it contains all its limit points. It is called compact if it is closed and bounded.*
   Let $\|\,.\,\|$ *be a vector norm on* $\mathbb{C}^n$. *A closed ball* B *of center* x^* *and radius* r *is defined by* $B = \{x \in \mathbb{C}^n,\ \|x - x^*\| \leq r\}.$

DEFINITION A.10   *A vector* x *is said to be an interior point of a set* A *if there exists some* $\varepsilon > 0$ *such that* $\{y \in \mathbb{C}^n,\ \|x - y\| < \varepsilon\} \subset A.$

DEFINITION A.11   *Consider a metric space* E *equipped with a distance* d. *A sequence* $\left\{x^{(k)}\right\}_{k\in\mathbb{N}}$ *of* E *is called a Cauchy sequence if for every* $\varepsilon > 0$, *there exists some* K *such that* $d(x^{(k)}, x^{(l)}) < \varepsilon$ *for all* $k, l \geq K$.

DEFINITION A.12   *A metric space in which every Cauchy sequence converges is called complete.*

DEFINITION A.13   *A Banach space is a complete normed vector space.*

References

[1] Omniorb web page. http://omniorb.sourceforge.net.

[2] P. R. Amestoy, I. S. Duff, and J.-Y. L'Excellent. Multifrontal parallel distributed symmetric and unsymmetric solvers. *Comput. Methods in Appl. Mech. Eng.*, 184:501–520, 2000.

[3] P. R. Amestoy, I. S. Duff, J.-Y. L'Excellent, and X. S. Li. Analysis and comparison of two general sparse sol vers for distributed memory computers. *ACM Transactions on Mathematical Software*, 27(4):388–421, 2001.

[4] P. R. Amestoy, A. Guermouche, J.-Y. L'Excellent, and S. Pralet. Hybrid scheduling for the parallel solution of linear systems. *Parallel Computing*, 32(2):136–156, 2006.

[5] J. Arnal, V. Migallon, and J. Penadès. Non-stationary parallel Newton iterative methods for nonlinear problems. *Lecture Notes in Comput. Sci.*, 1573:142–155, 1999.

[6] J. Arnal, V. Migallon, and J. Penadès. Parallel Newton two-stage multisplitting iterative methods for nonlinear systems. *BIT Num. Math.*, 43:849–861, 2003.

[7] O. Aumage, G. Mercier, and R. Namyst. MPICH/Madeleine: a True Multi-Protocol MPI for High-Performance Networks. In *Proc. 15th International Parallel and Distributed Processing Symposium (IPDPS 2001)*, page 51, San Francisco, April 2001. IEEE.

[8] O. Axelsson. A generalized SSOR. *BIT*, 13:443–467, 1972.

[9] O. Axelsson. Incomplete block matrix factorization preconditioning methods. The ultimate answer? *J. Comp. Appl. Math.*, 12&13:3–18, 1985.

[10] O. Axelsson. A general incomplete block-matrix factorization method. *Lin. Alg. Appl.*, 74:179–190, 1986.

[11] O. Axelsson. *Iterative Solution Methods*. Cambridge Univ. Press, Cambridge, 1994.

[12] J. Bahi. Asynchronous iterative algorithms for nonexpansive linear systems. *J. Parallel Distrib. Comput.*, 60(1):92–112, 2000.

[13] J. Bahi, S. Contassot-Vivier, and R. Couturier. Asynchronism for itera-
      tive algorithms in a global computing environment. In *The 16th Annual
      International Symposium on High Performance Computing Systems and
      Applications (HPCS'2002)*, pages 90–97, Moncton, Canada, June 2002.

[14] J. Bahi, S. Contassot-Vivier, and R. Couturier. Dynamic load balanc-
      ing and efficient load estimators for asynchronous iterative algorithms.
      IEEE Transactions on Parallel and Distributed Systems, 16(4):289–299,
      2005.

[15] J. Bahi, S. Contassot-Vivier, and R. Couturier. Evaluation of the asyn-
      chronous iterative algorithms in the context of distant heterogeneous
      clusters. *Parallel Computing*, 31(5):439–461, 2005.

[16] J. Bahi, S. Contassot-Vivier, and R. Couturier. Performance compari-
      son of parallel programming environments for implementing AIAC algo-
      rithms. *Journal of Supercomputing. Special Issue on Performance Mod-
      elling and Evaluation of Parallel and Distributed Systems*, 35(3):227–
      244, 2006.

[17] J. Bahi, S. Contassot-Vivier, R. Couturier, and F. Vernier. A decentral-
      ized convergence detection algorithm for asynchronous parallel iterative
      algorithms. *IEEE Transactions on Parallel and Distributed Systems*,
      16(1):4–13, 2005.

[18] J. Bahi and R. Couturier. Parallelization of direct algorithms using
      multisplitting methods in grid environments. In *19th IEEE and ACM
      Int. Parallel and Distributed Processing Symposium, IPDPS 2005*, pages
      254b, 8 pages, Denver, Colorado, USA, April 2005. IEEE Computer
      Society Press.

[19] J. Bahi, R. Couturier, D. Laiymani, and K. Mazouzi.   Java and
      asynchronous iterative applications: large scale experiments.   In
      *IPDPS'2007, 21th IEEE and ACM Int. Symposium on Parallel and
      Distributed Processing Symposium*, page 195 (8 pages), Long Beach,
      California USA, March 2007. IEEE Computer Society Press.

[20] J. Bahi, R. Couturier, K. Mazouzi, and M. Salomon. Synchronous and
      asynchronous solution of a 3d transport model in a grid computing
      environment. *Applied Mathematical Modelling*, 30(7):616–628, 2006.

[21] J. Bahi, R. Couturier, and P. Vuillemin. Solving nonlinear wave equa-
      tions in the grid computing environment: an experimental study. *Jour-
      nal of Computational Acoustics*, 14(1):113–130, 2006.

[22] J. Bahi, S. Domas, and K. Mazouzi. Combination of java and asyn-
      chronism for the grid : a comparative study based on a parallel power
      method. In *18th IEEE and ACM Int. Conf. on Parallel and Distributed
      Processing Symposium, IPDPS 2004*, pages 158a, 8 pages, Santa Fe,
      USA, April 2004. IEEE Computer Society Press.

[23] J. Bahi, S. Domas, and K. Mazouzi. Jace : a java environment for distributed asynchronous iterative computations. In *12th Euromicro Conference on Parallel, Distributed and Network based Processing, PDP'04*, pages 350–357, Coruna, Spain, February 2004. IEEE Computer Society Press.

[24] J. Bahi, S. Domas, and K. Mazouzi. More on jace: New functionalities, new experiments. In *IPDPS'2006, 20th IEEE and ACM Int. Symposium on Parallel and Distributed Processing Symposium*, pages 231–239, Rhodes Island, Greece, April 2006. IEEE Computer Society Press.

[25] J. Bahi, E. Griepentrog, and J. C. Miellou. Parallel treatment of a class of differential-algebraic systems. *SIAM Journal on Numerical Analysis*, 33(5):1969–1980, October 1996.

[26] J. Bahi and J.-C. Miellou. Contractive mappings with maximum norms. Comparison of constants of contraction and application to asynchronous iterations. *Parallel Computing*, 19:511–523, 1993.

[27] J. Bahi, J.-C. Miellou, and K. Rhofir. Asynchronous multisplitting methods for nonlinear fixed point problems. *Numerical Algorithms*, 15:315–345, 1997.

[28] R. E. Bank and C. C. Douglas. An efficient implementation of the SSOR and ILU preconditionings. *Appl. Numer. Math.*, 1:489–492, 1985.

[29] R. Barrett, M. Berry, T. F. Chan, J. Demmel, J. Donato, J. Dongarra, V. Eijkhout, R. Pozo, C. Romine, and H. Van der Vorst. *Templates for the Solution of Linear Systems: Building Blocks for Iterative Methods, 2nd Edition*. SIAM, Philadelphia, PA, 1994.

[30] G. M. Baudet. Asynchronous iterative methods for multiprocessors. *J. ACM*, 25:226–244, 1978.

[31] A. Berman and R. J. Plemmons. *Nonnegative Matrices in the Mathematical Sciences*. Academic Press, New York, 1979. Reprinted by SIAM, Philadelphia, 1994.

[32] D. Bertsekas. Distributed asynchronous computation of fixed points. *Math. Programming*, 27:107–120, 1983.

[33] D. P. Bertsekas and J. N. Tsitsiklis. *Parallel and Distributed Computation: Numerical Methods*. Prentice Hall, Engelwood Cliffs, 1989.

[34] D. A. Bini. Numerical computation of polynomial zeros by means of Alberth's method. *Numerical Algorithms*, 13:179–200, 1996.

[35] R. Bolze, F. Cappello, E. Caron, M. Daydé, F. Desprez, E. Jeannot, Y. Jégou, S. Lanteri, J. Leduc, N. Melab, G. Mornet, R. Namyst, P. Primet, B. Quetier, O. Richard, E.-G. Talbi, and I. Touche. Grid'5000: A large scale and highly reconfigurable experimental grid

testbed. *International Journal of High Performance Computing Applications*, 20(4):481–494, 2006.

[36] R. Chandra, L. Dagum, D. Kohr, D. Maydan, J. McDonald, and R. Menon. *Parallel Programming in OpenMP*. Morgan Kaufmann Publishers Inc., 2001.

[37] A. S. Charão. *Multiprogrammation parallèle générique des méthodes de décomposition de domaine*. PhD thesis, Institut National Polytechnique de Grenoble, 2001.

[38] D. Chazan and W. Miranker. Chaotic relaxation. *Linear Algebra Appl.*, 2:199–222, 1969.

[39] M. Cosnard and P. Fraigniaud. Analysis of asynchronous polynomial root finding methods on a distributed memory multicomputer. *IEEE Transaction on Parallel and Distributed Systems*, 5(6):639–648, 1994.

[40] R. Couturier and S. Domas. CRAC: a grid environment to solve scientific applications with asynchronous iterative algorithms. In *21th IEEE and ACM Int. Symposium on Parallel and Distributed Processing Symposium, IPDPS'2007*, page 289 (8 pages), Long Beach, USA, March 2007. IEEE Computer Society Press.

[41] H. Dag. An approximate inverse preconditioner and its implementation for conjugate gradient method. *Parallel Computing*, 33(2), 2007.

[42] T. Davis. University of Florida sparse matrix collection. NA Digest, 1997. See http://www.cise.ufl.edu/research/sparse/matrices/.

[43] E. W. Dijkstra, W. H. J. Feijen, and A. J. M VanGasteren. Derivation of a termination detection algorithm for distributed computation. *Information Processing Letters*, 16(5):217–219, 1983.

[44] J. Dongarra, A. Lumsdaine, X. Niu, R. Pozo, and K. Remington. A sparse matrix library in C++ for high performance architectures. In *Second Object Oriented Numerics Conference*, pages 214–218, 1994.

[45] I. S. Duff. A survey of sparse matrix research. In *Proceedings of the IEEE*, volume 65, pages 500–535, 1977.

[46] I. S. Duff, A. M. Erisman, and J. K. Reid. *Direct Methods for Sparse Matrices*. Oxford University Press, 1989.

[47] R. Duncan. A survey of parallel computer architectures. *IEEE Computer*, pages 5–16, Feb. 1990.

[48] D. El Baz. A method of terminating asynchronous iterative algorithms on message passing systems. *Parallel Algorithms and Algorithms*, 9:153–158, 1996.

[49] D. El Baz, P. Spitéri, J.-C. Miellou, and D. Gazen. Asynchronous iterative algorithms with flexible communication for non linear network problems. *Journal of Parallel and Distributed Computing*, 38:1–15, 1996.

[50] M. N. El Tarazi. *Contraction et Ordre Partiel Pour L'étude D'algorithmes Synchrones et Asynchrones En Analyse Numérique*. PhD thesis, Univ. de Franche-Comté, 1981.

[51] M. N. El Tarazi. Some convergence results for asynchronous algorithms. *Numer. Math.*, 39:325–340, 1982.

[52] D.J. Evans. The use of preconditioning in iterative methods for solving linear equations with symmetric positive definite matrices. *J. Internat. Math. Appl.*, 4:295–314, 1967.

[53] M. Fiedler and V. Ptak. On matrices with non-positive off-diagonal elements and positive principal minors. *Czechoslovak Math. J.*, 87:382–400, 1962.

[54] M. J. Flynn. Some computer organizations and their effectiveness. *IEEE Transactions on Computers*, C-21(9):948–960, September 1972.

[55] N. Francez. Distributed termination. *ACM Transactions on Programming Languages and Systems*, 2(1):42–55, January 1980.

[56] V. Frayssé, L. Giraud, S. Gratton, and J. Langou. Algorithm 842: A set of GMRES routines for real and complex arithmetics on high performance computers. *ACM Transactions on Mathematical Software*, 31(2):228–238, June 2005.

[57] R. W. Freund and N. M. Nachtigal. QMR: a quasi-minimal residual method for non-Hermitian linear systems. In *Iterative Methods in Linear Algebra*, pages 151–154. Elsevier Science Publishers, 1992.

[58] G. Frobenius. über matrizenaus positiven elementen. *S.-B. Preuss. Akad. Wiss., Berlin*, pages 456–477, 1912.

[59] A. Frommer. Parallel nonlinear multisplitting methods. *Numerische Mathematik*, 56:269–282, 1989.

[60] A. Frommer and G. Mayer. Convergence of relaxed parallel multisplitting methods. *Linear Algebra Appl.*, 119:141–152, 1989.

[61] A. Frommer and G. Mayer. On the theory and practice of multisplitting mehods in parallel computation. *Computing*, 49:63–74, 1992.

[62] A. Frommer and H. Schwandt. A unified representation and theory of algebraic additive Schwarz and multisplitting methods. *SIAM Journal on Matrix Analysis and Applications*, 18, 1997.

[63] A. Frommer and D. B. Szyld. Asynchronous iterations with flexible communication for linear systems. *Calculateurs Parallèles, Réseaux et Systèmes Répartis*, 10:421–429, 1998.

[64] A. Frommer and D. B. Szyld. On asynchronous iterations. *J. Comput. and Appl. Math.*, 123:201–216, 2000.

[65] A. Frommer and D.B. Szyld. Asynchronous iterations with flexible communications for linear systems. *Calculateurs Parallèles*, 10:421–429, 1998.

[66] A. Geist, A. Beguelin, J. Dongarra, W. Jiang, R. Manchek, and V. Sunderam. *PVM: A Users' Guide and Tutorial for Networked Parallel Computing*. MIT Press, 1994.

[67] A. V. Gerbessiotis. Architecture independent parallel binomial tree option price valuations. *Parallel Computing*, 30(2):301–316, 2004.

[68] L. Giraud and S. Gratton. A set of GMRES routines for real and complex arithmetics. Technical report, Cerfacs, 1998.

[69] A. Greenbaum, M. Rozložnik, and Z. Strakoš. Numerical behaviour of the modified Gram-Schmidt GMRES implementation. *BIT*, 37:706–719, 1997.

[70] L. Grigori and X. S. Li. A new scheduling algorithm for parallel sparse LU factorization with static pivoting. In *Super Computing 2002*. IEEE Computer Society Press and ACM Sigarch, 2002. Paper 139 on CD.

[71] W. Gropp, E. Lusk, and A. Skjellum. *Using MPI: Portable Parallel Programming with the Message Passing Interface*. MIT Press, 1994.

[72] W. Hackbusch. *Iterative Solution of Large Sparse Systems of Equations*. Springer, 1994.

[73] A. C. Hindmarsh and R. Serban. Example program for cvode. http://www.llnl.gov/CASC/sundials/.

[74] R. Hitchens. *Java NIO*. O'Reilly & Associates, Inc., 2002.

[75] M. Jones and D. B. Szyld. Two-stage multisplitting methods with overlapping blocks. *Numer. Linear Algebra Appl.*, 3:113–124, 1996.

[76] W. Kahan. *Gauss-Seidel Methods for Solving Large Systems of Linear Equations*. PhD thesis, University of Toronto, 1958.

[77] L. V. Kantorovich. Functional analysis and applied mathematics. *UMN*, 23, No. 6, 1948.

[78] L. V. Kantorovich. On Newton's method. In *Trudy Mat. Inst. Steklov 28*. 1949.

[79] M. A. Krasnosel'ski, G. M. Vainikko, P. P. Zabreiko, Y. B. Rutitskii, and V. Y. Stetsenko. Translated from the Russian by D. Louvish. *Approximate Solution of Operators Equations*. Wolters-Noordhoff Publishing, Groningen, 1972.

[80] H. T. Kung. Synchronized and asynchronous algorithms for multiprocessors. In *J. F. Traub Ed., Algorithm and Complexity: New Directions and Recent Results*, New York, 1976. Academic Press.

[81] C.-C. J. Kuo and B. C. Levy. A two–level four–color SOR method. *SIAM J. Numer. Anal.*, 26:129–151, 1989.

[82] X. S. Li and J. W. Demmel. SuperLU_DIST: A scalable distributed-memory sparse direct solver for unsymmetric linear systems. *ACM Transactions on Mathematical Software*, 29(2):110–140, June 2003.

[83] N. Lynch. *Distributed Algorithms*. Morgan Kaufmann, San Francisco, 1996.

[84] N. Maillard, E-M. Daoudi, P. Manneback, and J-L. Roch. Contrôle amorti des synchronisations pour le test d'arrêt des méthodes itératives. In *Renpar 14*, pages 177–182, Hamamet, Tunisie, April 2002.

[85] T. A. Manteuffel. An incomplete factorization technique for positive definite linear systems. *Mathematics of Computation*, 34:473–497, 1980.

[86] J.-C. Miellou. Algorithmes de relaxation chaotique à retards. *R.A.I.R.O.*, R-1:55–82, 1975.

[87] J.-C. Miellou, P. Cortey-Dumont, and M. Boulbrachene. Perturbation of fixed point iterative methods. *Advances in Parallel Computing*, 1:81–142, 1990.

[88] J. C. Miellou, D. El Baz, and P. Spitéri. A new class of asynchronous iterative algorithms with order intervals. *Math. of Computation*, 221(67):237–255, 1998.

[89] R. Namyst and J.-F. Méhaut. PM^2: Parallel multithreaded machine. A computing environment for distributed architectures. In *Parallel Computing: State-of-the-Art and Perspectives, ParCo'95*, volume 11, pages 279–285. Elsevier, North-Holland, 1996.

[90] O. Nevanlinna. Remarks on Picard-Lindelof iteration. *Bit*, 29:Part I, 328–346, Part II 535–562, 1989.

[91] N. K. Nichols. On the convergence of two-stage iterative processes for solving linear equations. *Siam J. Numer. Anal.*, 10:460–469, 1973.

[92] D. P. O'Leary and R. E. White. Multisplittings of matrices and parallel solution of linear systems. *SIAM J. on Alg. Disc. Math.*, 6:630–640, 1985.

[93] J. M. Ortega and W. C. Rheinboldt. *Iterative Solution of Nonlinear Equations in Several Variables.* Academic Press, New York, 1970.

[94] A. Ostrowski. über die determinanten mit überweigender hauptdiagonale. *Coment. Math. Helv.*, 10:69–96, 1937.

[95] A. Ostrowski. *Solution of Equations and Systems of Equations.* Academic Press, New York, 1966.

[96] B. Parhami. *Introduction to Parallel Processing - Algorithms and Architectures.* Plenum Series in Computer Science. Springer, 1999.

[97] O. Perron. Zur theorie der matrizen. *Math. Ann.*, 64:248–263, 1907.

[98] A. Pope. *The CORBA Reference Guide: Understanding the Common Object Request Broker Architecture.* Addison-Wesley, Reading, MA, USA, December 1997.

[99] B. Pugh and J. Spaccol. MPJava: High Performance Message Passing in Java using Java.nio. In *Proceedings of the Workshop on Languages and Compilers for Parallel Computing*, College Station, Texas, USA, October 2003.

[100] S. P. Rana. A distributed solution to the distributed termination problem. *Information Processing Letters*, 17:43–46, July 1983.

[101] F. Robert, M. Charnay, and F. Musy. Itérations chaotiques série parallèle pour des équations non-linéaires de point fixe. *Appl. Math.*, 20:1–38, 1975.

[102] Y. Saad. *Iterative Methods for Sparse Linear Systems.* PWS Publishing, New York, 1996.

[103] Y. Saad and M. Schultz. GMRES: A generalized minimal residual algorithm for solving nonsymmetric linear systems. *SIAM Journal of Scientific and Statistical Computing*, 7:856–869, 1986.

[104] S. A. Savari and D. P. Bertsekas. Finite termination of asynchronous iterative algorithms. *Parallel Computing*, 22:39–56, 1996.

[105] J. G. Siek and A. Lumsdaine. The matrix template library: A generic programming approach to high performance numerical linear algebra. In *ISCOPE*, pages 59–70, 1998.

[106] B. Smith, P. Bjorstad, and W. Gropp. *Domain Decomposition.* Cambridge University Press, Cambridge, 1996.

[107] A. Van Der Steen and J. Dongarra. Overview of recent supercomputers. http://www.top500.org/orsc/2006/, 2006.

[108] P. Stein and R.L. Rosenberg. On the solution of linear simultaneous equations by iteration. *J. of London Math. Soc.*, 23:111–118, 1948.

[109] D. B. Szyld and M. Jones. Two-stage and multisplitting methods for the solution of linear systems. *SIAM J. Matrix Anal. Appl.*, 13:671–679, 1992.

[110] M. El Tarazi. Some convergence results for asynchronous algorithms. *Numer. Math.*, 39:325–340, 1982.

[111] A. Uresin and M. Dubois. Sufficient conditions for the convergence of asynchronous iterations. *Parallel Computing*, 10:83–92, 1989.

[112] H. A. van der Vorst. *Preconditioning by Incomplete Decompositions*. PhD thesis, University of Utrecht, 1982.

[113] R. S. Varga. *Matrix Iterative Analysis*. Prentice Hall, 1962.

[114] R. S. Varga. On recurring theorems on diagonal dominance. *Linear Algebra Appl.*, 13:1–9, 1976.

[115] H. F. Walker. Implementation of the GMRES method using Householder transformations. *SIAM Journal on Scientific and Statistical Computing*, 9(1):152–163, 1988.

[116] J. K. White and A. E. Sangiovanni-Vincentelli. *Relaxation Techniques for the Simulation on VLSI Circuits*. Kluwer Academic Publishers, Boston, 1987.

[117] R. E. White. Parallel algorithms for nonlinear problems. *SIAM J. Alg. Discrete Meth.*, 7:137–149, 1986.

[118] R. E. White. Multisplittings with different weighting schemes. *SIAM J. Matrix Anal. Appl.*, 10:481–493, 1989.

[119] P. Wolfe. *Methods of nonlinear programming*. John Wiley, New York, USA, 1976.

[120] D. M. Young. On the accelerated SSOR method for solving large linear systems. *Advances in Mathematics*, 23(3):215–271, 1977.

[121] J. Zhang. A sparse approximate inverse preconditioner for parallel preconditioning of general sparse matrices. *Applied Mathematics and Computation*, 130(1):63–85, July 2002.

Index

A

AIAC *64*, 112, 132
application granularity 186
asynchronism xvi
    model 113
    total 118

B

Banach space 116, *204*
binomial tree 179

C

Cauchy sequence 204
chaotic relaxation 112
closed set 204
compact set 204
complete space 204
computational units xv
contraction *5*, 104
    approximate 7
convergence xvi
    conditions *7*, 116
    detection *104*, 145
    speed *13*, 33
COW 56
CRAC 133, *180*

D

derivative 40
    partial 41
descent methods 25
differential equation 8
direct algorithms . . . . xvi, *12*
discretization
    space *8*, 196
    time 197
distributed cluster 57
distributed memory machine 54

E

eigenvalue 2
eigenvector 2
error vector 14

F

fixed point 6
    equation 12
    mapping 14

G

Gauss-Seidel algorithm 12, *17*
Gauss principle 112
GMRES 27
gradient method 25
    biconjugate 32
    conjugate *26*, 88
gradient of a function 41
grid 57
GRID'5000 186

H

heterogeneity xv
hierarchical architecture 60
homogeneity 50

I

ILU factorization 38
interior point 204
iterative algorithms . . . . *xvi*, 12
    block versions *20*, 71, 88
    comparison *14*, 17, 23, 74
    linear *12*, 71
    nonstationary . . . *12*, 23, 81, 117
    parallel . . . . *see* PIA
    sequential 6
    stationary 12
    two-stage *21*, 117

J

JACE 133, *177*
Jacobi algorithm 12, *15*, 85, 135

L

Lanczos biorthogonalization......32
leader election protocol 147
limit point.....................204
linear systems..............*11*, 186
Lipschitz continuity..............43
local cluster 56

M

mapping
    contractive...................5
    fixed point 14
        extended.................76
    linear 1
    nonlinear.................5, *40*
matrix
    H-matrix 82, *202*
    M-matrix ... 119, 125, 188, *202*
    Z-matrix..................202
    (ir)reducible 201
    bandwidth 188
    block diagonal 76
    block tridiagonal 22
    condition number 27
    diagonal 15
    diagonalizable...............32
    diagonal dominance 201
    Hermitian...................1
    Hessenberg..................29
    identity 2
    inverse 1
    invertible....................1
    iteration............. *13*, 17, 19
    Jacobian 41
    nonsingular ...... *see* invertible
    positive definite............202
    positive semidefinite.......201
    preconditioning 33
    symmetric...................1
message passing.........57, *82*, 132
meta-cluster xv

MIMD...........................52
minimization algorithms 24
MISD 51
MPICH/Madeleine 174
multisplitting......xvii, *75*, 91, 120,
    138, 193
    two-stage..... *79*, 111, 138, 142
multisplitting-Newton 131, 142,
    197
multithreaded environments....*133*,
    174

N

Newton-Jacobi...................79
Newton-multisplitting *80*, 101,
    129, 140, 197
Newton algorithm 41
nonlinear systems...........*39*, 196
norm
    A-norm 24
    contraction................104
    Euclidean *2*, 105
    Frobenius 4
    matrix 3
    max 105
    maximum 3
    residual 29
    vectorial 2
NOW..............................56

O

O'Leary and White algorithms...77
OmniOrb 4.....................174
overlapping strategies .. *96*, 138, 193
overrelaxation algorithm 12, *19*

P

parallelism xv
    coarse grained..... *83*, 111, 186
    fine grained 82
parallel architectures.............51
parallel machine 52
parallel programming.......xvi, 82,
    173
Perron-Frobenius theorem 203

PIA 61
PM2 174
preconditioning *33*, 82
projection methods 24
pseudo-periods 157
PU 52

R
relaxation parameter............. 19
residual.................... *105*, 146
    vector...................... 26
Richardson algorithm 12

S
scalar product 24

Schwarz algorithms *78*, 127
shared memory machine...... *53*, 84
SIAC 63
SIMD 51
SISC 62
SISD 51
SOR........................... 19
spectral radius................... 2
SSOR...................... 35–38
steering....................... 114
stopping criterion *7*, 147, 158

T
transposes 1